For what can war but endless war still breed?
Till truth and right from violence be freed.

—Milton, Sonnet to Fairfax[1]

The only question is whether this modern intensification of the military element is, upon the whole, salutary for the interests of humanity or otherwise—a question, which it would be about as easy to answer as the question of war itself—we leave both to philosophers.

—Clausewitz, On War[2]

1 Milton, John. "Sonnet 15: Fairfax." *The Sonnets of John Milton.* Legare Street Press, 2022.
2 Clausewitz, Carl. *On War.* Princeton University Press, 1989, chapter 26.

Pacifism as War Abolitionism

Responding to the unprecedented violence of our times, and the corresponding interest in nonviolent solutions, this book takes up the heart of pacifism: its critique of what pacifists have termed the *war system*.

Pacifism as War Abolitionism provides an account of the war system that draws on contemporary sociology, history, and political philosophy. The core of its critique of that system is that war begets war, and hence war will not be ended—or even constrained—by finding more principled ways to fight war, as many imagine. War can only be ended by ending the war system, which can only be done *nonviolently*. This has been the message of pacifism's great voices like Gandhi, Dr. Martin Luther King, Jr., and Dorothy Day. It is the principal message of this book.

Key Features

- Draws extensively on the sociological and historical research on war to expand the usual philosophical discussion beyond hypothetical accounts
- Expands the dialogues on the ethics of war beyond *just war theory* to its principal alternative: *pacifism*
- Engages discussion of empire and imperialism in relation to the logic and development of the war system
- Presents pacifism's response to the reality of war today, including the idea of "never-ending war"

Cheyney Ryan is a senior research fellow at Oxford University's Institute for Ethics, Law, and Armed Conflict, where he focuses on nonviolence, pacifism, and the critique of just war theory. He has been named one of the leading scholars in peace and conflict studies by the *Washington Post* and received the Joseph J. Blau Prize from the Society for Advancement of American Philosophy for significant contributions to the history of American philosophy. He is co-chair of the Oxford Consortium for Human Rights, and has received wide recognition for his work on social justice, including an Honorary Doctor of Humane Letters from Quinnipiac University for his "steadfast commitment to peace on our planet."

Pacifism as War Abolitionism

Cheyney Ryan

Routledge
Taylor & Francis Group

NEW YORK AND LONDON

Designed cover image: © Getty Images

First published 2024
by Routledge
605 Third Avenue, New York, NY 10158

and by Routledge
4 Park Square, Milton Park, Abingdon, Oxon, OX14 4RN

Routledge is an imprint of the Taylor & Francis Group, an informa business

© 2024 Cheyney Ryan

The right of Cheyney Ryan to be identified as author of this
work has been asserted in accordance with sections 77 and 78
of the Copyright, Designs and Patents Act 1988.

ISBN: 978-1-032-68617-2 (hbk)
ISBN: 978-1-032-68614-1 (pbk)
ISBN: 978-1-032-68618-9 (ebk)

DOI: 10.4324/9781032686189

Typeset in Sabon
by Apex CoVantage, LLC

Contents

Preface: "Lay Down Your Arms!"   *viii*

1   Grand Illusions   1

2   Pacifism as Tradition   42

3   Personal and Political Pacifism   81

4   The Dynamic of the War System   122

5   Disarming Power: Memory, Sacrifice, Grief, and Hope   163

Appendix: Pacifism and Self-Defense   182
Bibliography   194
Index   210

Preface: "Lay Down Your Arms!"

What do we mean by "pacifism"?

A premise of this work is that pacifism is ill-conceived as a fixed "position" that can be quickly summarized and easily assessed (and for some, easily dismissed). Pacifism is best understood as a tradition and a movement; more specifically, as a tradition of *dialogue* about how best to abolish war and a movement of *activism* committed to achieving that end. In this regard, it is no different from any other political orientation that we take seriously. "Liberalism," for example, names both an ongoing dialogue around shared values and an ongoing project for realizing those values. Taking pacifism seriously for me begins with taking seriously the thinkers and activists that have committed themselves to it, including their serious disagreements about why and how to end war.

Why take pacifism seriously?

At one time, pacifists were the only ones who believed that war should end and could end. The rise of what I shall call war system engendered a counter-reaction so that by the end of the 19th century and up to our own time many others have envisioned ending war as their goal. For some, this has meant promoting "wars to end all war"—the "War on Terror" is a recent iteration of this. For others, it has meant promoting arrangements to significantly constrain war, if not end it entirely, through armed enforcement, on the assumption that ultimately violence only respects violence.

There are many reasons for taking pacifism seriously. A starting point is the fact that the alternative approaches to addressing the scourge of war have quite obviously failed. No sooner had World War II ended than the world found itself in a Cold War; the optimism at the end of the Cold War quickly gave way to the "War on Terror," the "War on Terror" quickly gave way to a new "Great Power Conflict"—and so on. It is possible that war cannot be ended. If so, humanity's future is dark indeed. There is no certainty that pacifism's critique of the war is correct, nor any guarantee that war can be ended by nonviolence. But if we can be confident of anything, it is that war will not be ended by more war. Being serious about ending war means taking pacifism seriously.

Why take abolishing war seriously?

War is not the only challenge facing humanity today. It may not even be the most serious challenge. But war looms over all of them if they are to be resolved. Its ultimate threat of nuclear conflict remains very much with us. Daniel Deudney has noted that, to find an historical analogy for nuclear conflict, one must imagine the great cataclysms like the fall of Rome, the Mongol invasions, the Black Plague, the European invasion of the Americas, and the two world wars "occurring at once and greatly compressed in time, perhaps into a single afternoon."[3] A theme of this work is that much of what goes on as war is more out of sight and insidious. Nuclear cataclysm may or may not happen. But as I write this, radioactive sludge from the Hanford nuclear reservation is slowly creeping toward my home in Oregon, silently but inexorably. (I discuss this in Chapter 5.) In 2023, global military expenditures reached an all-time high of $2,240 billion. This has doubled in real terms in the last 20 years. The United States accounts for 40% of global spending, which is three times more than China the second largest spender. NATO countries account for more than 50% of all military spending. A Scientific American report states that if countries reduced their spending by just 2% a year, the money saved could end global poverty and fund the climate goals set by developed countries.[4] If you read at the normal rate of 250 words a minute, this means that, in the three minutes or so it has taken you to read this, the world has spent $12,785,388 on war.

The title of this Preface, "Lay Down Your Arms!", is an homage to a founding voice of modern pacifism, Bertha von Suttner (1843–1914) and her international bestseller *Die Waffen nieder!* which helped alert her generation to the threat of war. Drawing on the rich tradition of pacifism, my book explores how to critique and confront the war system. It argues that political philosophers should adopt a more expansive approach to thinking about war generally, one that draws on political sociology and history as well as the normative discussions of these matters.

---

3 Deudney, Daniel. "Nuclear Weapons and the Waning of the Real-State." *Daedalus*, 124(2), 1995, p. 210.
4 Rovelli, Carlo and Smerlak Matteo. "A Small Cut in World Military Spending Could Help Fund Climate, Health and Poverty Solutions." *Scientific American*, March 17, 2022, https://www.scientificamerican.com/article/a-small-cut-in-world-military-spending-could-help-fund-climate-health-and-poverty-solutions/.

[I]/ Where I Am Coming From

My approach to these questions is deeply informed by my own dialogues and activism over the years which deserve mention at the start.

I have been involved with pacifism for a long time. I was introduced to nonviolence by the 1960s civil rights movement. My first political experience was attending the 1963 March on Washington with my parents. It was the Vietnam War that started me thinking about pacifism when I decided to register as a conscientious objector to the draft at age 18. At the time, conscientious objector status was only granted for opposition to *all* war. Opposition to a *particular* war like Vietnam was not sufficient. This is when I started thinking seriously about what it meant to be unconditionally opposed to all war. My views were deeply influenced by my parents, Jessica Cadwalader Ryan and Robert Ryan. My thinking about how pacifism might inform one's whole life was profoundly impacted by my involvement with the Catholic Worker Movement and its founder Dorothy Day. Dorothy is now recognized as one of pacifism's major voices. Her religious-inflected pacifism is not the focus of this work. But I remain deeply attracted to it, which is why I take the religious perspective seriously when I do speak to it, and more generally why I do not shy away from invoking religious/spiritual frameworks where appropriate.

I continued to be involved in antiwar activism through the Vietnam War and afterward, including the anti-nuclear/anti-interventionist movements of the 1980s and opposition to the "War on Terror." But questions of political violence were not always central to my academic work. My PhD dissertation was on the philosophy of economics. While my early publications included works on pacifism, I wrote on other topics as well. My eventual focus on pacifism actually resulted from taking *war* more seriously. Several things prompted this. One was encountering the historical literature that started appearing in the 1990s on why the 20th century had been so warlike. I think the first was Gabriel Kolko's *Century of War: Politics, Conflict, and Society Since 1914* (1994), followed by Eric Hobsbawm's *Age of Extremes: The Short Twentieth Century 1914–1991* (1995) and Mark Mazower's *Dark Continent: Europe's Twentieth Century* (2000).[5] They made a huge impression on me. Despite my long involvement in antiwar work, I had never thought of my era this way. I also encountered the literature in political sociology on the formative role of war in the origins and development of states and empires. Another factor was the post-9/11 recognition that, contrary to the widespread belief of the 1990s, war was not going

---

5 Kolko, Gabriel. *Century of War: Politics, Conflicts, and Society Since 1914*. New Press, 1994; Hobsbawm, Eric. *Age of Extremes: The Short Twentieth Century, 1914–1991*. Abacus, 1995; Mazower, Mark. *Dark Continent: Europe's Twentieth Century*. Vintage, 2000.

away, on the contrary, the question was whether the 21st century would be "another century of war."[6]

Finally, there was the deepening—and discouraging—awareness that political philosophy did not take war seriously at all. Jonathan Glover began his important work *Humanity: A Moral History of the 20th Century* by citing R. G. Collingwood's observation, "The chief business of 20th century philosophy is to reckon with 20th century history."[7] But reckoning with the wars of the 20th century has been far from philosophy's chief business. This began to change with the flourishing of just war theory after 9/11. I shall discuss how this has contributed to the robust discussions of pacifism now occurring. But war remains a marginal issue for political philosophers, as I would illustrate with this anecdote. Some years ago, I gave a symposium paper at the American Philosophical Association on war's relation to the state and its general importance for political philosophy. I was pleased to have had it accepted. I was dismayed to see that when the conference program arrived my talk was categorized as "Military Ethics." This would be like categorizing an essay on the state's right to punish as "police ethics." But it made sense insofar as today's political arrangements seek to insulate the citizenry from the state's war making endeavors, such that the average person now has little idea of where their country is fighting or what it is fighting for.

There are parallels to this. Following the American Civil War, American political philosophers forgot about that conflict and the issues it had raised about slavery in particular and racism in general. This was part and parcel of the white power structure's relegating it to the margins. In the United States, I think something of the same sort happened with the forgetting of the Vietnam War and the issues it had raised about the centrality of war to American politics. This is why my argument for taking pacifism seriously is as much an argument for taking war seriously.

[II]/ Who I Am Indebted to

Many people have impacted my views on pacifism and war. I am grateful for the companionship of David Frank and Stephen Stern in thinking about and working for social justice over the years. I benefited from discussions with David Richie, Deen Chatterjee, Rob Gould, Betty Reardon, Lani Roberts, Leslie Scott, and Johanna Luttrell in Oregon. Moving to Oxford meant becoming part of the wider just war community from which

6 Kolko, Gabriel. *Another Century of War?* New Press, 2004.
7 Glover, Jonathan. *Humanity: A Moral History of the 20th Century.* Yale University Press, 2001, p. v.

I continue to learn. This began with my involvement with Oxford's Changing Character of War Program (CCW) which first alerted me to the problem of discussing "war" in the abstract given its dramatic changes over time. It also provoked my interest in how to tell the longer story of war in ways that would provide a critical perspective on its past and future. The program's director Sir Hew Strachan was supportive throughout.

My greatest debt is to David Rodin and Henry Shue, who invited me to join Oxford's Institute on Ethics, Law, and Armed Conflict (ELAC) upon its founding. Their own writings are a model of philosophical clarity and political seriousness to which we should all aspire. Their friendship and personal support for me have been a great blessing. Moving to Oxford also led to my involvement with Merton College, which has been a home away from home for me and my wife Sandy at Oxford. Among the many who enriched my thinking about war and peace, I give special thanks to Cecile Fabre, Helen Frowe, Jeff McMahan, Seth Lazar, Bob Underwood, and Mike Robillard, mindful of the significant—though amiable—disagreements that remain between us. I thank Andrew Beck and Marc Stratton at Routledge for their support and patience, and the anonymous reviewers of the manuscript.

Hugo Slim has been a constant source of insights about humanitarianism and war and an endless source of inspiration in our efforts together on the Oxford Consortium for Human Rights. Sujata Gadkar-Wilcox, Katie Dwyer, Johanna Luttrell, Ashleigh Landau, and Tamara Niella have also been invaluable partners in the Consortium, where I have had wonderful interchanges over the years with the young people from all over the world who have attended our workshops. Ken Pendleton has been a great help in preparing the manuscript.

My aforementioned political activities have contributed to a rocky academic career at times. This has led to some special debts. After I was expelled from Harvard for antiwar activities, Hilary Putnam single-handedly saved my academic career by contacting Alasdair MacIntyre and Marx Wartofsky and urging them to take me on at Boston University. A question that many of us faced was the value of pursuing academic work versus the other ways one could contribute to social justice. It was Howard Zinn, my other graduate school advisor, who persuaded me by his words and example that academics and activism could be joined. I dedicate this book to his memory.

Dr. Sandy Stein Ryan and I met on a picket line supporting striking hospital workers in the 1970s. Over the years, in conjunction with her own writings on the marginalization of women in art and her own teaching as a professor of art, we have discussed the myriad ways in which social justice can be pursued. Her wisdom, guidance, and support have been central to every aspect of my life over the decades. She exemplifies the virtues of truth and compassion at the heart of the project to peace. I have been blessed that she has shared her life with me.

This book draws on previously published writings of mine, included in the Bibliography. The basic framework is one that I first sketched in the article on pacifism for the *Oxford Handbook on the Ethics of War*.[8] It was further developed in an article on the pacifist critique of just war theory for *The Routledge Handbook of Pacifism and Nonviolence* and in an article on pacifism and terrorism in the collection *Pacifism's Appeal Ethos, History, Politics*.[9] My discussion of the ethics of personal pacifism is foreshadowed in my early article, "Pacifism, Self Defense, and the Possibility of Killing" and developed further in "The Morning Stars Will Sing Together: Compassion, Nonviolence, and the Revolution of the Heart" in *The Routledge Handbook of Love in Philosophy*.[10] My discussion of war and nationalism draws on " 'Wretched Nurseries of Unceasing Discord': War, Nationalism, and the Politics of Peace" that appeared in Theoretical Inquiries in Law (2020).[11]

I have benefited from a great many discussions in many places about pacifism over the years. Some of this material was presented as the Tanner-McMurrin Lecture at Westminster College, Utah, the "Or" Emet Lecture at the York University's Nathanson Center on Transnational Human Rights, the Leon Peters Ethics Lecture at California State University, Fresno, and the Walter Powell Lectures at Linfield College, Oregon. I have had productive exchanges from the presentations I have given at symposia on pacifism in Antwerp, Belgium, Berlin, Germany, New South Wales, Australia, Belgrade, Serbia, Stockholm, Sweden, Oslo, Norway, and Philadelphia and Nashville, USA. Other talks/discussions on pacifism and just war theory have been given at the Tel Aviv University Faculty of Law, Australian National University, the British Political Science Association in Belfast, Northern Ireland, and the philosophy departments at Cambridge University, Manchester University, University of Birmingham, Freie Universitat Berlin, University of Pavia, University of Aarhus, University of

8 "Pacifism." *Oxford Handbook on the Ethics of War*. Edited by Seth Lazar and Helen Frowe, Oxford University Press, 2016.

9 "The Pacifist Critique of Just War Theory." *The Routledge Handbook of Pacifism and Nonviolence*. Edited by Andrew Fiala (2017). "War, Hostilities, Terrorism: A Pacifist Perspective." *Pacifism's Appeal Ethos, History, Politics*. Edited by J. Kustermans, T. Sauer, D. Lootens and B. Segaert, 2018.

10 "Pacifism, Self Defense, and the Possibility of Killing." Ethics, Jan. 1983. "The Morning Stars Will Sing Together: Compassion, Nonviolence, and the Revolution of the Heart." *The Routledge Handbook of Love in Philosophy*. Edited by Adrienne Martin, Routledge, 2019.

11 "Wretched Nurseries of Unceasing Discord": War, Nationalism, and the Politics of Peace." *Theoretical Inquiries in Law*, 2020.

Copenhagen, University of Southern California, University of Michigan, University of Utah, University of Maryland, Gettysburg College, Willamette University, University of California, Riverside, and meetings of the American Philosophical Association, the Radical Philosophy Association, the North American Society for Social Philosophy, and Concerned Philosophers for Peace. Last but not least have been the many presentations/discussions over the years at Oxford University for CCW and ELAC as well as for OxPeace, Balliol College, St. John's College, and New College.

Chapter 1

Grand Illusions

As war is the system of Government on the old construction, the animosity which Nations reciprocally entertain, is nothing more than what the policy of their Governments excites to keep up the spirit of the system. Each Government accuses the other of perfidy, intrigue, and ambition, as a means of heating the imagination of their respective Nations, and incensing them to hostilities. Man is not the enemy of man, but through the medium of a false system of Government. Instead, therefore, of exclaiming against the ambition of Kings, the exclamation should be directed against the principle of such Governments; and instead of seeking to reform the individual, the wisdom of a Nation should apply itself to reform the system.

—Thomas Paine, *Rights of Man*

When people hear the word "pacifism," they imagine a world in which great numbers of people are rendered helpless before the predations of aggressive forces, a world in which too many people live in permanent insecurity if not in fear for their very survival, a world in which innocent men, women, and children are periodically slaughtered by political tyrants with impunity.

We have good reason to fear such a world—since this is exactly what centuries of "just" wars have given us. Just war thinking claims to tell us what "good wars" look like. A larger perspective suggests that "good" wars produce "bad" ones produce "good" ones, and so on because they are all part of a larger system for which justice and injustice are essentially irrelevant. War's many "justifications" come and go, ever changing and constantly contradicting each other. Yet war continues—with little prospect of escaping this cycle unless our fundamental categories are rethought and our basic practices revised.

[I]/ Our Violent Age

We approach our topic in the shadow of an exceptionally violent age.

How violent?

DOI: 10.4324/9781032686189-1

[1]/ The Worst Century—Ever?

Writing at the end of the 20th century, historian Eric Hobsbawm observed in his *Age of Extremes*, "During the short century more human beings had been killed or allowed to die by human decision than ever before in history." It was without doubt "the most murderous century of which we have record," not only due to "the scale, frequency and length of the warfare which filled it, barely ceasing for a moment in the 1920's" but also due to the "unparalleled scale of the human catastrophes" such warfare produced, from "the greatest famines in history to systematic genocide."[1] He cites Isaiah Berlin: "I have lived through most of the 20th-century without, I must add, suffering personal hardship. I remember it only as the most terrible century in Western history."[2] Sir Martin Gilbert began his three-volume history of the century "The 20th-century has witnessed some of humanity's greatest achievements and some of its worst excesses." He cited Winston Churchill, "It is called the century of the common man because in it the common man has suffered most."[3]

Here are the figures behind such judgments.

From 1900 to 1990 (Hobsbawm's "short 20th century"), the world saw 237 new wars whose battles killed approximately one million persons per year. By the century's end, that number climbed to approximately 275 wars and 115 million deaths in battle. While averages are misleading, since most of the deaths occurred in the two world wars, this comes to about 3,150 deaths per day or about 130 deaths per hour, 24 hours a day throughout the century.[4] And this does not include civilian deaths—which are harder to estimate, though for the first time in history civilian deaths in the century due to war outnumbered those of soldiers. Estimates are that at the beginning of the 20th-century military personnel accounted for 90% of the casualties of war, while at the end of the century civilians accounted for that same percentage.[5] Total deaths due to war in the 20th century could have approached 250 million, or almost 7,000 people a day, 300 people per hour. No other century in history approaches these totals in number of conflicts and people killed. The comparative numbers are equally striking. Consider the percentage of people killed relative to the total population.

1   Hobsbawm, E. J. *Age of Extremes: The Short Twentieth Century, 1914–1991*. Abacus, 1995, p. 12.
2   *Ibid.*, p. 1.
3   Gilbert, Martin. *A History of the Twentieth Century 1900–1933*, vol. 1. William Morrow and Company, 1997, pp. 1–2.
4   Tilly, Charles. *Coercion, Capital, and European States: AD 990–1992*. Blackwell, 2015, p. 67.
5   See Chesterman, Simon. *Civilians in War*. Lynne Rienner Publishers, 2001.

The 18th century saw five deaths per thousand people due to war, the 19th century six deaths per thousand, and the 20th century 46 deaths per thousand—that is, almost *eight* times higher than the previous century.

From 1480 to 1800, a significant new conflict started every two to three years, in the 19th century every one to two years, since then every 14 months. Looming over them all has been the 20th century's two world wars. As Niall Ferguson writes,

> Even allowing for the accelerating growth in the world's population, the world wars were the most destructive in history. Somewhere in the region of 2.4% of the world's entire population was killed in the Second World War . . . compared with roughly .4% in the Thirty Years War and .2% in the Napoleonic Wars and the War of the Spanish Succession. The total death toll in the First World War amounted to something like 1 percent of the pre-war population of all 14 combat countries, 4% of all males between 15 and 49 and 13% of all those mobilized. . . . In the Second World War roughly 3% of the entire prewar population of all combatant countries died as a result of the war.[6]

Casualties have declined since World War II, which Ferguson has attributed to the fact that wars "have generally been fought against far less well-equipped opposition." "The world has not become that much more peaceful," he has written, "It is just that the overwhelming majority of the victims of war have been Asians or Africans."[7] According to one estimate, the total war-induced death toll for 1945–1999 was more than 20 million. The Cold War saw an average of more than 1,200 people (most of them civilians) dying in war of one type or another, every day, for 45 years. This equaled more than three My Lai massacres every day for 45 years.[8]

The 21st century has witnessed America's post-9/11 wars in Afghanistan and Iraq. Their cost has been substantial. Almost a million people have died in the post-9/11 wars due to direct war violence, and at least 4.5 million people have died from the reverberating effects of war in Afghanistan, Iraq, Pakistan, Syria, Yemen, Libya, and Somalia, a significant proportion

---

6 Ferguson, Niall. *The Cash Nexus: Economics and Politics from the Age of Warfare through the Age of Welfare, 1700–2000.* Basic Books, 2002, pp. 34–35.

7 Ferguson, *Cash Nexus*, p. 36.

8 Walter LaFeber cites the number of 21 million deaths for the Cold War in LaFeber, Walter. "An End to Which Cold War." *The End of the Cold War: Its Meaning and Implications.* Edited by Michael J. Hogan, Cambridge University Press, 1992, p. 13. See also Chamberlin, Paul Thomas. *The Cold War's Killing Fields: Rethinking the Long Peace.* Harper Collins, 2019.

of them children.[9] From 2018 to 2020, the United States government undertook what it labeled "counter-terrorism" activities in 85 countries. So far, the deadliest conflict of the 21st century has been outside Europe. The Second Congo War (1998–2003) has also been called the "Great War in Africa" or "Africa's First World War." By 2008, it had caused 5.4 million deaths, mainly civilians, with another two million displaced. This and other wars occurred largely outside of the West's awareness.

But will war return to the West?

If so, this would repeat the morbid logic that I describe in this work—of war periodically migrating away from Europe only to return in a more virulent form. At the 20th century's end, the West declared a resounding victory in the Cold War. The nuclear standoff had been diffused; the superiority of Western civilization had been conclusively demonstrated; former enemies would now work together as friends. All the risks and costs of the Cold War would be justified by the era of enduring peace to come. What kind of "victory" lands the world in yet another "Cold War" just three decades later? As I write this, the War in Ukraine (2014–) appears to be the Korean War (1950–1953) *redux*. It is raising the same talk of "existential threats" and "defending ultimate values" from both sides. It is posing the same questions about "who-provoked-who" (note that while the United States takes full credit for ending the last Cold War, it takes zero responsibility for starting the new one). Ukraine seems destined to become a Korea-type stalemate that will simmer for decades. All we can say for certain is that the crisis is the occasion for massive military buildups and Manichaean ideologies of the world divided between "adversaries" and "friends." The main difference is that now the United States has a new and potentially bigger "adversary" in China.[10] The question is not one of "good guys" versus "bad guys," anymore than the endless conflicts of the last century could reduced this. It is a question about a larger system of war in which all of us are implicated.

Jonathan Glover's aforementioned *Humanity: A Moral History of the 20th Century* remains the best discussion by a political philosopher of the challenge posed by war though it appeared more than two decades ago. "It is a myth that barbarism is unique to the 20th-century," he writes. "The whole of human history includes wars, massacres, and every kind of torture and cruelty. But it is still right that much of 20th-century history has been a very unpleasant surprise."[11] "Unpleasant surprise," indeed—there's

---

9 "How Death Outlives War: The Reverberating Impact of the Post-9/11 Wars on Human Health," report of Watson Institute Costs of War Project, May 2023.

10 See Cave, Damien, Pierson, David and Buckley, Chris. "China and U.S. Lay Out Rival Visions for Asia as Ships Nearly Collide." *New York Times*, June 4, 2023.

11 Glover, Jonathan. *Humanity: A Moral History of the 20th Century.* Yale University Press, 2001, pp. 5–6.

British understatement for you. The continuing violence is such that historians now debate whether we have witnessed another Thirty Years War in the 20th century or are still witnessing another Hundred Years War persisting into the 21st century. Steven Pinker has suggested that overall violence has declined nevertheless. The scholarly community remains unconvinced by his argument.[12] It may be that the average person is more secure than in past times if one sets war aside. But this just confirms that the problem is not violence generally, but *war*, a problem that if anything is highlighted by society's success at achieving security in other realms.

[2]/ Political Philosophy in a Time of Crisis

The past century or so has been a time of deep crisis. It is not unlike the 17th century in this regard, when Europe was torn apart by the holy wars of religious factions killing each other in the name of God—leading to the recognition of many that political thinking and practice needed to be profoundly reoriented.

At that time, political philosophy faced two choices.

One was to develop a *better* theory of religious war: to specify more precisely and more compellingly the conditions in which persons could kill others and sacrifice themselves for their faith. The other choice was to question the whole idea of "religious war" and of the social arrangements that produced it, to imagine, that is, alternative ways of constructing social arrangements, including new ways of conceiving power. At the time, the first choice was overwhelmingly the more "reasonable" one. After all, religious-based polities and their religious conflicts had been fixtures for millennia, to the point of being regarded as an essential part of the "natural" order. Moreover, applying a more sophisticated "religious-war theory" to the continuing conflicts—"What are the pro's and con's of the Bohemian Revolt? How far does liability extend for the Scanian War?"— would ensure that experts in such a theory would be listened to in the halls of power, that they would gain prominent positions in the 17th century's equivalent of today's "think tanks," and that they would be interviewed on the 17th century's equivalent of "cable news."

But deep problems require new—and more radical—solutions. This was the assumption of those who envisioned and promoted an entirely different order of the type we now identify with modernity. It meant reassessing the very nature of political power including the individual's place in its arrangements and how it all related to the fundamental logic of war itself: why do we have war in the first place, when is it truly necessary—indeed, what makes war necessary at all? With no promise of success, those who

---

12 See Mann, Michael. *On Wars*. Yale University Press, 2003, pp. 271–311. Braumoeller, Bear. *Only the Dead: The Persistence of War in the Modern Age*. Oxford University Press, 2019.

asked such questions laid the basis for the liberal democratic arrangements we now cherish. But experience now shows that whatever else such arrangements have achieved, they will not end war.

I think we stand at a similar juncture today. After another century of catastrophic wars and the crises they engendered, one response would be to tinker with existing ideas about war and existing conceptions of the societies that wage them. This too is the overwhelmingly "reasonable" approach. But more radical solutions are required. Ultimately, critics of "religious wars" did not reject war *per se;* rather, they replaced it with secular notions of "just war" and a new conception of the place of war in the society of states then emerging. The upshot was that waging war all but replaced promoting religion as what the state was principally about, and modernity became a story of constant warfare. Just as the notion of "religious wars" was once rejected, I think it is time for the notion of "just wars" to be rejected. Doing this is not just a matter of tinkering. As before, it means imagining and promoting an entirely different order from the type we know now.

"Pacifism" is the name for this radical project of rethinking. In a violence-besotted world like ours, where war is baked into our institutions and mindsets, pacifism invariably appears unreasonable. There is certainly no guarantee of its success. So it is best regarded as a project of hope for a world in which the continuing resort to violence offers no hope.

This book is an invitation to join this project.

[3]/ What This Project Involves

Pacifism is often dismissed as an unrealistic doctrine, one that is hopelessly detached from the way the world really is. My view is the opposite. I take pacifism as the most *realistic* response to the persistent violence of our times. This is why I began with today's violence and shall return to it shortly.

Put simply, pacifism is the new realism. This dictates my approach to the topic.

Recent decades have witnessed a robust revival in thinking about "just wars." But its approach has been abstract in two respects. It has proceeded *hypothetically*, for example, it has talked about "states," "wars," etc., with no attention to the specific institutions or practices involved, especially how they have arisen and changed in history. Or, when it attends to actually existing war, it has proceeded *anecdotally*, that is, it has cherry-picked examples of war from across the decades or centuries or millennia, with no attention to context. It is hard to imagine any other social practice addressed in this abstract way. No one would propose an ethic of, say, the family which assumed there was one thing—"the" "family"—stable across history and independent of all social contexts.

The problem is exacerbated by the extent to which political philosophy generally has been a demilitarized zone. Its dominant approaches have almost nothing to say about war as a central concern of societies and a formative influence on societies. This is not to belittle the issues that political philosophy does address. But the fact remains that no one reading the major works of recent decades would have any idea that they were written in the shadow of one of the most warlike eras in history.

The upshot has been that political philosophers' discussions of war have proceeded independently of the significant developments in the study of war in such fields as political sociology and history. I proceed differently. Much of my discussion will draw on this literature, with special attention to the extensive discussions of war's relation to the central institutions of our time—states, nations, and empires. This means that my discussion may be demanding one for those unfamiliar with this literature. But there is no other way to make a case against war than to discuss actually existing war, what it has been and what it currently is. My hope is that it can serve as an introduction to this work in other fields and an inspiration to move beyond mere "intuitions" about war to engaging war in its empirical and historical reality.

[II]/ Pacifism as War Abolitionism

Pacifism is a tradition of activism, argument, and reflection that has meant different things.[13] Any account of it will be as much a proposal as a description. I shall take the core of pacifism to be *war abolitionism*. This is how other major studies of pacifism have conceived it.[14] However much

---

13 On pacifism as tradition of argument, see Cortright, David. *Peace: A History of Movements and Ideas*. Cambridge University Press, 2008, p. 30. For the pacifist tradition, see Brock, Peter. *Pacifism in Europe to 1914*. Princeton University Press, 1972; Brock, Peter. *Freedom from Violence: Sectarian Nonresistance from the Middle Ages to the Great War*. University of Toronto Press, 1991; and Brock, Peter and Young, Nigel. *Pacifism in the 20th Century*. Syracuse University Press, 1999.

14 See, for example, Teichman, Jenny. *Pacifism and the Just War*. Basil Blackwell, 1986, pp. 3–4; Norman, Richard. *Ethics, Killing and War*. Cambridge University Press, 1995, p. 197; Cady, Duane. *From Warism to Pacifism: A Moral Continuum*. 2nd ed., Temple University Press, 2010. Fiala, Andrew. *Transformative Pacifism: Critical Theory and Practice*. Bloomsbury, 2018. Jackson, Richard. "Bringing Pacifism Back into International Relations." *Social Alternatives*, 33(4), 2014, pp. 63–66. Dobos, Ned. *Ethics, Security, and the War-Machine: The True Cost of the Military*. Oxford University Press, 2020. The *Oxford English Dictionary* defines pacifism as "the doctrine or belief that it is desirable and possible to settle international disputes by peaceful means." The *Routledge Handbook of Pacifism and Nonviolence* begins by noting that "pacifism is typically used in a narrow sense to mean opposition to war, while nonviolence typically describes a method and a means, and in some cases a virtue or even a way of life" (ed. by Andrew Fiala, Routledge, 2018, p. 1.)

pacifists disagree on other things, they agree on the necessity of abolishing war and achieving peace. So that is my focus here.

At the heart of war abolitionism is its critique of and confrontation with the *war system*. The core of that critique is the claim that *war begets war*: war will not be ended, or even constrained, by finding more effective or more principled ways to fight it. This has only bred more war (in Milton's words). So, the core of its message is that war can only be ended by ending the war system—and this can only be done *nonviolently*. Nonviolence is not just a preferable way to solving human problems; it is the *only* way to solve them if we are ever to escape the war system and replace it with a peace system that can ensure a truly human future.

I take this to be the message of pacifism's great voices like Gandhi, Dr. Martin Luther King, Jr., Dorothy Day, and others. It is the principal claim of this book.

[1]/ Critiquing and Confronting the War System

The notion of a "war system" has played a prominent role in pacifist and antiwar thinking since at least the 18th century. Thomas Paine's remarks are typical. "War" is not just an action or event; it is a "system of Government" whose "spirit" is pervaded by conflicts that "heat the imagination" and "incense" societies to "hostilities." "Man is not the enemy of man" in any fundamental sense; war is not the product of any human nature. Rather, they are made enemies through a "false system of Government"—his example is the British Empire—so our aim must be not to reform the individual but to fundamentally change the system. We find this stress on war as a system throughout the tradition. The 19th-century pacifist and anti-slavery activist Charles Sumner wrote in his "The War System and the Commonwealth of Nations" (1849) that "[a]s Slavery is an Institution, growing out of local custom, sanctioned, defined, and established by Municipal Law, so War is an Institution, growing out of general custom, sanctioned, defined, and established by the Law of Nations." The key institutions involve preparing for war, or what I call war building. "The component parts of the War System," he wrote, "may all be resolved into Preparations for War. Hence, only when we contemplate War in this light can we fully perceive its combined folly and wickedness."[15] The great 20th-century antiwar voice Randolph Bourne wrote in "War is the Health of the State" (1918), "War is a function of this system of States, and could not occur except in such a system."[16]

---

15 Sumner, Charles. *An Address Before the American Peace Society*. American Peace Society, 1854, p. 11.
16 Bourne, Randolph. *War Is the Health of the* State. Anecdota Press, 2015, p. 18.

It has featured prominently in numerous contemporary figures such as Andrew Alexandra, Duane Cady, Richard Falk, Joshua Goldstein, Robert Holmes, Mary Kaldor, Betty Reardon, Martin Shaw, and Jonathan Schell, among others. I take a major contribution to this work to unpack the structure and history of the war system in ways that may inspire others to critique it and confront it.[17]

As I construe it, speaking of war as a "system" implies several key features:

1. War possesses an autonomous logic:

War intertwines with other social factors. To speak of war as a "system" is to say that its impact is pervasive. In recent years, theorists of militarism have written extensively about war's sometimes subtle and insidious impact on all dimensions of life including the most personal ones. Significant work has been done on the relation of militarism to gender and to race. Scholars like Judith Butler have addressed how violence is implicit in our very notions of subjecthood. A full account of the war system would include these features.[18]

My account will especially focus on the relation between political and economic factors. This is a matter of emphasis as much as anything else, as well as a reflection of my own scholarly background.

But to say that war's logic is an autonomous one is to say that it is not *reducible* to anything else. It is not a manifestation of "deeper" economic factors, or "deeper" cultural factors, or "deeper" psychological factors, or "deeper" anything else. It is not an epiphenomenon of capitalist relations,

---

17 Alexandra, Andrew. "Political Pacifism." *Social Theory and Practice*, 29(4), 2003, pp. 589–606; Cady, Duane. *From Warism to Pacifism: A Moral Continuum*. 2nd ed., Temple University Press, 2010; Falk, Richard and Kim, Samuel S. *The War System: An Interdisciplinary Approach*. Routledge, 2020; Goldstein, Joshua. *War and Gender: How Gender Shapes the War System and Vice Versa*. Cambridge University Press, 2009; Holmes, Robert. *Pacifism: A Philosophy of Nonviolence*. Bloomsbury Academic, 2017; Kaldor, Mary. *New and Old Wars: Organized Violence in a Global Era*. Polity Press, 2012; Reardon, Betty. *Sexism and the War System*. Syracuse University Press, 1996; Shaw, Martin. *Dialectics of War: An Essay in the Social Theory of Total War and Peace*. Pluto Press, 1988; Schell, Jonathan. *The Unconquerable World: Power, Nonviolence, and the Will of the People*. Penguin, 2005.

18 See, for example, Lutz, Catherine. "Making War at Home in the United States: Militarization and the Current Crisis." *American Anthropologist*, 104(3), pp. 723–735; Enloe, Cynthia. *Globalization and Militarism: Feminists Make the Link*. Rowman & Littlefield Publishers, 2016; Singh, Nikhil Pal. *Race and America's Long War*. University of California Press, 2019; Butler, Judith. *The Force of Non-Violence: An Ethico-Political Bind*. Verso, 2021. Cawston, Amanda. "'Shorn of All Camouflage': Re-assessing the Problem of Violence." Doctoral Thesis, Cambridge University, 2014.

as Marxists would have it—though it stands in a complex relation to capitalism; nor is it an epiphenomenon of undemocratic institutions, as some liberals would have it—though it stands in a complex relation to democracy.

This is why I insist that ending war is not a matter of the undoing some *further* factor that causes war (perhaps by waging war against *it*). Ending war means undoing war's own autonomous logic.

2. *Its autonomous logic is an inexorable one.*

"Inexorability" was Randolph Bourne's term for how war takes on a life of its own that seems to transcend all human agency. War is both inevitable and merciless. This has been a recurring theme in the critique of war. Others have invoked metaphors of war as a "self-perpetuating machine" or "Frankenstein monster." Sociologist C. Wright Mills spoke of the "fearful symmetry" to describe the relentless logic by which states constantly provoked each other into conflict.[19] For Bourne, its inexorability meant that any attempts to "manage" war in good directions rather than bad ones rested on a fundamental misunderstanding of war's logic. War can be neither constrained nor ended by tinkering with it, but only by abolishing the war system *in toto*.

Its inexorability is grounded in the fundamental logic of war. That logic is *twofold*: war is both an act of violence and an exercise of power. It is the latter that distinguishes war from other forms of collective violence. As Clausewitz, the first great theorist of war noted, war is violence in service of power, or more simply: war is *political killing*. And it is its political nature that renders war *promiscuous*, in serving something—power—whose concentration and expansion means it increasingly takes on a life of its own. As Clausewitz put it, the purpose of war may be to serve social purposes; the nature of war is to serve only itself.

This points to a fundamental difference between pacifism and just war thinking. The latter presumes an *instrumentalist* picture of war. War is to be appraised in terms of *why* that instrument is employed (Does it abide by the principles of *jus ad bellum*?) and *how* that instrument is employed (Does it abide by the principles of *jus in bello*?). It is a picture of war that views it as primarily a matter of violence. Pacifism claims that once the dimension of power is introduced, this instrumentalist picture must be abandoned. War is not an instrument that we employ; we are an instrument that war employs, and thus its inexorability.

---

19 Mills, C. Wright. *The Causes of World War Three*. Ballantine Books, 1958, p. 9.

We shall see how the critique of predatory power has been a persistent feature of pacifism in its various forms. It reflects the extent to which pacifism in its various forms has arisen in response to *empire*: to the Roman Empire, with early Christian pacifism; and to European empires, with the more political pacifisms of the 19th and 20th centuries. So its concern has been with the kind of illicit power that empire represents. This should frame our understanding of nonviolence. Nonviolence is not just the refusal to engage in the killing of war. It aims to fashion a new *form* of power that gives full expression to human agency rather than rendering such agency a tool for power's own autonomous purposes. I shall speak of it as a *disarming power*: a power that disarms armed power by being disarming—in ways that do not provoke a further armed response.

3. War's inexorability is grounded in the institutions of war building.

Modern war has been an enterprise of states, empires, and nations. But understanding war is not just a matter of what those institutions *do*. It is more fundamentally a matter of what they *are*. They are artifacts and enablers of war, specifically of what I term *war building*. Historically, states, empires, and nations prevailed over other political forms by their success at mobilizing the human and material resources for war—to the point that their very *identities* are constituted by war.

Thus, war's inexorability reflects how it is baked into the fundamental institutions of modernity. So ending the war system means fundamentally altering these institutions and the identities that have defined them.

4. And the logic of war possesses its own historical dynamic:

That war is a system means it possesses a history, which I claim follows a certain pattern—specifically, its long-term arc has been a *cyclical* one. Here, a parallel with Marx's critique of capitalism is illuminating. Marx regarded capitalism as conjuring powers of production that transcended all control, thus leading to periodic and ever-deepening economic crises. Similarly, I regard the war system as conjuring powers of destruction that transcend all control, thus leading to periodic and ever-deepening political crises.

Indeed, I suggest that each crisis has had the form of "the chickens coming home to roost." This means that in each case war, having been displaced from the European core to the non-European periphery, has returned to the European world with a vengeance. Socialists have claimed the choice is ultimately between socialism and barbarism. I suggest that the choice is ultimately between pacifism and barbarism. The future will depend on how seriously we take the challenge of war.

5. Finally, the logic of war is one of contradiction.

Jonathan Schell in *Unconquerable World* speaks of a "peace system" arising alongside the war system. My discussion draws on what Charles Tilly has termed the central "paradox" of war and the European state. I term it the *Paradox of European Modernity*. In Tilly's words, it means that the liberal arrangements of the state have been a "by-product" of the state's pursuit of war and struggles to expand its military capacity.[20] In the 20th century, total war's need to engage the whole society meant that political and economic benefits were demanded in return. Scholars of economic inequality have argued that the *only* progress in economic fairness has been the result of war and to a lesser extent some of the revolutions that war engendered. In *The Great Leveler: Violence and the History of Inequality from the Stone Age to the Twenty-First Century*, Walter Scheidel claims that violent conflict, "served as a uniquely powerful catalyst for equalizing policy change, providing powerful impetus to franchise extensions, unionization, and the expansion of the welfare state."[21] Thomas Picketty writes,

> To a large extent, it was the chaos of war, with its attendant economic and political shocks, that reduced inequality in the twentieth century. There was no gradual, consensual, conflict-free evolution toward greater equality. In the twentieth century it was war, and not harmonious democratic or economic rationality, that erased the past and enabled society to begin anew with a clean slate.[22]

Hence, the story of war has been one of extremes. Physical destruction has been accompanied by material abundance; violent policies have been accompanied by expanding political and social rights—at least for some. As some might say, the logic of the war system has been a *dialectical* one.

But these have not just been offshoots of the war system; they have been reactions against and sites of resistance to the war system. This is crucial to my account of pacifism's origins. Pacifism is sometimes seen as an ultra-moralistic doctrine whose judgments are imposed from outside

20 Tilly, Charles. *Coercion, Capital and European States*, p. 206. See also Tarrow, Sidney. *War, States, and Contention: A Comparative Historical Study*. Cornell University Press, 2015; Tilly, Charles. "Where Do Rights Come From?" *Democracy, Revolution, and History*. Edited by Theda Skocpol, Cornell University Press, 2000, pp. 55–72.

21 Scheidel, Walter. *The Great Leveler: Violence and the History of Inequality from the Stone Age to the Twenty-First Century*. Princeton University Press, 2018, p. 7.

22 Piketty, Thomas. *Capital in the Twenty-First Century*. Translated by Arthur Goldhammer, Harvard University Press, 2014, p. 275.

social reality, as it were. This might fit some forms of religious pacifism. But the pacifism that most concerns me grounds its judgments in the war system itself, specifically by taking its counter tendencies to their logical conclusion. Its critique of the war system is grounded in the war system.

Some insist that the counter tendencies have triumphed. John Keegan argues that, while war has been the principal business of states in the past, the late 20th century saw their "civilianization" such that war in Western countries has been marginalized in favor of social benefits. A number of the social theorists that I draw on would maintain that, while my emphasis on war has been true of the past, it is now outdated. These matters are complex. But my own view is that to regard war as no longer integral to what states, nations, or empires are all about rests on an overly optimistic view about the West's turning away from war that contemporary developments like the war in Ukraine do not support. If what we have now is a "dual" civil-military state, the military part remains subtly dominant, while the civil part can be a source of resistance to it.

[2]/ What Kind of Position Is Pacifism?

The single greatest error in approaching pacifism is conceiving it as just one "policy position" among others that stands or falls on its judgment of particular wars. Hence, the quarrel with pacifism tends to be one of "What would you say about war X? . . . Or what about war Y?"

There are several problems with framing things this way (e.g., it presumes that wars can be easily individuated and assessed as such). But the main point is that pacifism's focus is more *fundamental* than this.

In *Arguing About War*, Michael Walzer construes pacifism as holding wars are never justified because they are always criminal acts. This is a common conception of pacifism, but it construes the critique too narrowly. Pacifism's focus is not the individual act of war but the larger system of war; its focus is not particular wars as criminal acts but war generally as a system of *organized crime*. This does not mean that the specifics of particular wars can just be ignored, especially when their particular justification seems strong. I shall speak to the problem of the particular case when I consider the unconditional character of pacifism's war condemnation.

There are different ways that the war-as-organized-crime model has been developed. The picture that I shall develop is of war as a form of *protection racket*. This is a picture of war that emerged in the 18th century. Rousseau wrote, "War furnishes a pretext for exactions of money, and another, no less plausible, for keeping large armies constantly on foot, to hold their people in awe." Charles Tilly pressed this point in his famous essay, "War Making and State Making as Organized Crime." "If protection rackets represent organized crime at its smoothest," he writes, "then war risking and state making—quintessential protection rackets with the advantage of legitimacy—qualify as our largest examples of organized crime."[23] Or, as two-time Congressional Medal of Honor winner Major General Smedley Butler titled his best-selling 1930's book, *War Is a Racket*. The system is maintained via a set of delusions, mainly that states are essential for "protecting" us from threats—while they themselves are the *source* of those threats. This has recently figured prominently in feminist critiques which have stressed the gendered dimension in ideologies of "protection."[24]

The Mafia-like dimension explains the obsession of politics with "security," as in today's preoccupation with "self-defense." It is evoked in that moment in "The Godfather" trilogy when Michael Corleone recounts to his daughter why his life has been one of crime, killing, and extortion. His explanation: "I wanted to protect you"—oblivious to the fact that his methods of protection were what prompted the need for "protection."[25] In 1962's Cuban Missile Crisis, the ideology of "national security" brought the world to within 15 minutes of nuclear holocaust. An ideology of self-protection that leads to the brink of collective suicide probably needs rethinking.

[3]/ No Slam-Dunk Arguments—or Solutions

Arguments about pacifism tend to conceive it as akin to *vegetarianism*, construed as the refusal to eat certain things. Hence, just as vegetarians are challenged with "Would you eat X? Would you eat Y?" pacifists are challenged with: "Would you kill in circumstances Z?" Just as one might parse different forms of vegetarianism in terms of what they would or would not eat, different forms of pacifism are parsed in terms of when they would or would not kill. This lends itself to bizarre hypotheticals.

23　Evans, Peter, et al. editors. *Bringing the State* Back. Cambridge University Press, 1985, p. 169.

24　Sjoberg, Laura. "Witnessing the Protection Racket: Rethinking Justice in/of Wars Through Gender Lenses." *International Politics*, 53(3), 2016, pp. 361–384. Sjoberg, Laura and Peet, Jessica. "A(nother) Dark Side of the Protection Racket." *International Feminist Journal of Politics*, 13(2), 2011, pp. 163–182. Young, Iris Marion, "The Logic of Masculinist Protection: Reflections on the Current Security State." *Signs: Journal of Women in Culture and Society*, 29(1), 2003, pp. 1–25. Kaldor, Mary. "Protective Security or Protection Rackets? War and Sovereignty." *Arguments for a Better World: Essays in Honor of Amartya Sen, Volume 2: Society, Institutions, and Development*. Edited by Kaushik Basu and Ravi Kanbur, Oxford University Press, 2008, pp. 470–487.

25　See Cavanaugh, William. "A Fire Strong Enough to Consume the House: The Wars of Religion and the Rise of the State." *Modern Theology*, 11(4), 1995, pp. 397–420. "The main difference between Uncle Sam and the Godfather is that the latter did not enjoy the peace of mind afforded by official government sanction" [p. 413].

Hypothetical arguments against pacifism ("Would you kill someone who is already about to die if a million lives could be saved by it?") are like hypothetical arguments against vegetarianism that ask: "Would you eat a hamburger if it was all that was left on earth and your doing so meant the survival of the human race?"

Pacifism is better thought of as akin to *feminism*. As I construe feminism, it is not the refusal to engage in particular acts so much as opposition to a general system, call it gender oppression. It is "gender-oppression abolitionism." And its impulse is not purely negative. It aims to articulate and nurture an alternative to the system of gender oppression. As such, feminism is best approached *as* a tradition, containing multiple voices, any account of which would begin with the story of those voices. Accordingly, I approach pacifism as a tradition containing multiple voices, the story of which will begin in the next chapter.

It follows that there are no slam-dunk arguments for or against pacifism, any more than there are ones for or against feminism. My own experience is that people often take there to be two slam-dunk arguments against pacifism. One is "What about self-defense?"; the other is "What about World War II?" *Qua* slam-dunks, they are often intended not to initiate serious discussion so much as to dismiss it. I shall have things to say about both questions, as I think the issues they raise are serious ones. But the fact is that neither side in the quarrel has said enough either pro- or con-war because neither side has sufficiently addressed the empirical reality of the war system to anything like the extent that, say, feminism has addressed the gender system.

Nor does pacifism offer any slam-dunk *solutions*, any more than feminism does. It has taken a very long time for systems of oppression like war, sexism, and racism to mess up the world, so fixing it is no simple matter. The critique of the war system is not without policy implications. I shall have much to say about the specific proposals that proponents of nonviolence have offered over time. But reducing pacifism to a set of policy proposals ignores the extent to which pacifism (like feminism) has been a *movement*—animated by a shared commitment to abolishing war but manifesting itself in different practical proposals. It also ignores the *radicalness* of the pacifist project, one that I have identified with a reinvisioning of politics generally, like that of previous revolutionary eras.

[III]/ The Challenges to Pacifism

So mine is an ambitious project, if not outrageously so. If war is truly such a constitutive feature of the modern world then unpacking its logic via the war system means providing a perspective on modernity itself. Any discussion of it will be abbreviated to the extreme, at best a crude summary of matters that could fill volumes.

But as someone once said: hopeless tasks are not always thankless! There is a place for Big Pictures, if only because we already carry Big Pictures in our head and questioning them starts by considering an alternative. So I shall paint in very broad strokes. To the usual flaws of such an endeavor, I shall add a few more of my own, nurtured over the years. One is my occasional lapses into polemics. My excuse is that if the problem of war is as urgent as I take it to be, then it warrants a certain degree of alarm. Less excusable, perhaps, are my occasional lapses into irony, if not sarcasm. I shall have occasion to remark on the dismissive attitude toward pacifism, especially over the last century. Serious criticisms need to be addressed. But what one mainly encounters are ill-informed simpleminded rejections of it. A lifetime of encountering such breezy dismissals has left me with a residue of annoyance that I do little to hide. One of my heroes, Henry David Thoreau, once mused that it is hard to criticize institutions that one regards as quite patently absurd. In the case of war, has any other social practice so consistently *failed*, indeed so consistently achieved its *opposite*? Yet those who reject it *in toto*—pacifists—are deemed the "unreasonable" ones!

Finally, mine is a Western-inflected account. I discuss the pacifism of other political/religious traditions in the next chapter but I mainly draw on the tradition I know best, mindful of the parochialism that this involves. It is also American-focused, but this has a bit more justification. The United States is currently the dominant military power in the world, and the story of modern war, at least since World War I, has been the story of its rise. Confronting war begins with the policies and practices of the United States, though it certainly does not end there.

There are several specific challenges to any discussion of pacifism worth noting at the start.

[1]/ Marginalized and Maligned

Pacifism has always been a marginal position. This is true of any radical position; indeed, it has been true of *every* political position when it first appeared. But pacifism has not just been marginalized, it has been *maligned*—dismissed by many as the paradigm of moral idiocy. Richard Jackson has portrayed it as a form of "subjugated knowledge."[26] The next chapter's review of the pacifist tradition will show that pacifism was once

---

26 Jackson, Richard. "Pacifism: The Anatomy of a Subjugated Knowledge." *Critical Studies on Security*, 6(2), 2018, pp. 160–175.

regarded with great seriousness, both theoretically and practically. The question I shall address is *why* it became so maligned in the 20th century.

But pacifists have not done nearly enough to state their case. This is changing with the flourishing in recent years of interest in pacifism, anti-militarism, and the critique of traditional justifications of war. Pacifism has been primarily a practical tradition whose principal audience has been the marginalized and whose main voices have been political activists, inordinately people of color and women. It has never had a significant presence in academia. For much of the 20th century, pacifists were barred from teaching in American public universities so they were mainly found in religious institutions. This is why the literature on religious pacifism has been so much more robust. I once lost a job for writing about pacifism. So theoretical work on it has been relatively undeveloped up to now, meaning that the project is still beginning.

[2]/ Disillusionment and Defection

Other challenges come from the opposite direction.

Anyone comparing a century ago to now might say that yesterday's pacifism has become today's common sense. We shall see how pacifism emerged at the start of the last century when many loudly celebrated war because they saw it as an intrinsically *good* thing—a theater of masculine virtue and a source of political regeneration. Such justifications were consistent with the racist, imperialist, aggressively nationalist mindsets of the time. American President Theodore Roosevelt praised war as "bully fun" and once urged that America invade Canada if no other war could be found. No one today thinks that war is "bully fun." A candidate who urged invading Canada would not be elected president; he or she would be deemed insane. John Mueller writes, "War is no longer regarded like eating or sex, as something inevitable to the human condition. It is seen as something grafted onto human existence, like dueling, which human beings can well do without."[27]

Let us term this the "*Grand Disillusionment.*"

Alongside it has been what might be termed the "*Great Defection.*"

I mean by this that the average citizen in Western countries has absolutely no interest in fighting in war him or herself. The best study of this by James J. Sheehan is aptly titled *Where Have All the Soldiers Gone?: The*

---

27 Mueller, John. "The Obsolescence of Major War." *Bulletin of Peace Proposals*, 21(3), 1990, p. 321.

Transformation of Modern Europe.[28] Sheehan observes that the disinterest of ordinary people to involve themselves in actual war making reflects disillusionment with war generally (though this has not meant the end either of war or of preparing for war).[29] With Europe, Sheehan suggests that while states, in his words, "have retained the capacity to make war with one another they have lost all interest in doing so." He cites a 2003 poll that asked if war was, "under certain circumstances, necessary to obtain justice"; only 12% in France and Germany agreed.[30] This meant that, judging by another poll, the percentage of French people believing in war was significantly less than those believing in flying saucers.[31]

The United States seemed to be the outlier here. The same 2003 poll reported that 55% of Americans "strongly agreed" that war was sometimes necessary to obtain justice. But the "War on Terror" took the bloom off that rose. No sooner had America invaded Afghanistan than its leaders discovered that, while most Americans supported war, hardly any wanted to fight in it. Enlistments declined dramatically, especially among African-Americans (down by 40%) and women (down by 23%) who were, in the words of one report, "marching away from offers to join the army."[32] Here, I refer readers to my 2009 polemic, *The Chickenhawk Syndrome: War, Sacrifice, and Personal Responsibility.*[33]

So the citizenry "celebrates troops" but no one wants to become one. The citizenry "honors service" while no one wants to serve. This has significantly altered the quarrel between the pacifist and the non-pacifist. It used to be that what distinguished pacifists from others was the question of fighting for one's country. A pacifist was someone who refused to do so. When I was a young man, when conscription was in force, what it *meant* to be a pacifist was that one would not serve in the military. Now, hardly anyone wants to serve in the military, much less fight in wars. It used to be that wholeheartedly supporting one's country's going to war

---

28 Sheehan, James L. *Where Have All the Soldiers Gone?: The Transformation of Modern Europe.* Houghton Mifflin, 2008.

29 See Vayrynen, Raimo, editor. *The Waning of Major War: Theories and Debates.* Routledge, 2005. This cultural shift, if enduring, means reassessing much of the literature on war, specifically on its so-called enduring attractions, insofar as a common theme has been that war persists because of its "excitement" to young men.

30 The Economist, June 5, 2004, cited Sheehan, p. xvi.

31 Bockman, Chris. "Why the French State Has a Team of UFO Hunters." *BBC News*, November 2014, https://www.bbc.com/news/magazine-29755919.

32 Bruns, Robert. "Blacks, Women Retreat from Army." *Associated Press*, March 9, 2005; Walsh, David. "Opposition to Iraq War Hitting US Military Recruitment." *World Socialist Web*, March 2005. www.wsws.org/en/articles/2005/03/mili-m12.html.

33 Ryan, Cheyney. *The Chickenhawk Syndrome: War, Sacrifice, and Personal Responsibility.* Roman and Littlefield, 2009.

meant seriously considering fighting in that war oneself—but no longer. This has lent a degree of unreality to academic discussions. When some academics insist that certain wars must be fought, they are not making a claim about what they *themselves* would do. Their claim is about what *other* people should do. It is like Catholic priests arguing about how many children families ought to have. It used to be that pacifists spent time urging people to resist serving in the military. Today, ordinary people need no such convincing.

[3]/ Why "Pacifism"?

The upshot is that it has become increasingly difficult to say what is *distinctive* about the pacifist position as opposed to one that is just deeply skeptical of war. What is the difference between pacifism and a fastidious just war approach, like that of today's Revisionists—many of whose views suggest that war is hardly ever justified? For some such thinkers, there *is* no sharp distinction. Pacifism is what you get from taking just war principles seriously.

Let us call this the *Demarcation Problem.*

The problem is raised by how I define pacifism. I have characterized it as war abolitionism, but pacifists have hardly been the *only* war abolitionists, if we construe this as a serious commitment to ending war. On the contrary, the horrors of the 20th century provoked significant efforts at *ending* war through international institutions and arrangements, which pacifists have supported.

So why not *forget* about the term "pacifism," especially given the stigma now surrounding it, and just speak of "war abolitionism"? Some have urged that pacifists do this.

We should not get hung up on labels. In my view, it is a reflection of our war-degraded culture that we lack adequate terms for these matters. Consider the fact that we do not even have a word for what it means to *not* be a pacifist! Duane Cady has suggested "war-ist," but I suspect that many would bridle at this. We shall see that pacifists have disagreed among themselves about what to call themselves. So let me explain why I remain committed to the term.

One is my conception of pacifism as both a tradition and movement, and I know of no other term that captures it in its full sweep. Moreover, I think that avoiding the term "pacifism," especially given what we shall see was its original expansive meaning, strikes me as surrendering to those whose maligning of the term has really been a maligning of peace generally. But there are more substantive reasons for preferring "pacifism" over a more general "war abolitionism."

For me, a pacifist is someone who *begins* with the question of why the other approaches to ending war have *failed*. I remarked on this in the Preface. What is distinctive about pacifism as war abolitionism is that it takes the failure of *other* war abolitionisms seriously. A parallel is the 19th-century movement against slavery. Radical abolitionists began with the question of why other anti-slavery movements had achieved so little. Their answer was twofold. They held that others had failed to appreciate the *depth* of the problem that slavery posed; specifically, that slavery constituted a social system with a logic of its own. And their failure to recognize this fact meant that others were too willing to compromise with the slavery system in ways that defeated their ultimate purposes. They needed a deeper theory and a more radical approach.

Accordingly, I take pacifism as war abolitionism as marked by two features.

One is its account of the war system, which I argue is distinguished from other forms of war abolitionism in its account of why we have war at all. War is not the product of some *further* factor: war is the product of war itself. From this, a second feature follows. I take pacifism's response to the failures of other approaches as motivating its *uncompromising* approach to both appraising and opposing war. War must always be condemned because its principal outcome is to engender more war; and it must always be opposed—which means not just opposing each war as it comes along but developing an alternative to war *per se* in robust arrangements of nonviolence.

[IV]/ The Spectacle of War

> We take pacifism seriously! But we must get our artillery budget through.
>
> General Stumm, in Robert Musil's
> *The Man Without Qualities* (1930–1942)

But people are not moved by arguments alone. Noting the widespread disillusionment with war at the end of the 20th century, John Keegan suggested that it was less a matter of "the calculation of political interest" than people's being moved by "repulsion from the spectacle of what war does."[34] Let me introduce this work and some of its main themes by describing how a pacifist might regard the spectacle of what war has been in our times.

34 Keegan, John. *A History of Warfare*. Alfred A. Knopf, 1997, p. 58.

[1]/ Optimism and Blindness

The 20th century began on a high note in Europe and the United States. It was an age of optimism and faith in progress. It is said to have been an age of innocence.

It was also an age of ignorance.

The optimism rested on indifference to events outside of Europe. In the 20th century's first genocide, 80% of South West Africa's Herero and Namia peoples perished between 1904 and 1907 as German colonists sought to "clear" them of their lands. Ten million people perished in the crimes of the rubber-forced labor system of the Belgian Congo. But the greatest case of blindness involved China's Taipang Rebellion (1850–1864) which, like other imperialist-induced conflicts, remains largely unknown in the West.

In a recurrent pattern of the war system, European powers recovered from their early 19th-century wars with each other by turning to foreign empire—and China. "Opening China's doors" to European trade promised massive new markets. The product that would lead the way was a favorite of the British East India Company: *opium*. But China refused to admit the drug, leading to one of the world's great narco-wars in which Britain was joined by other Western powers in humiliating China into allowing the narcotic and other Western goods. The upshot was disaster for the political and social fabric. It laid the basis for the rise of the charismatic figure Hong Xiuquan, who led a religious/nationalist revolt to overthrow the Qing dynasty and institute a new political/moral order. The ensuing conflict took between 20 million and 30 million lives, as many if not more than World War I, perhaps the deadliest civil war in history.[35]

Still, within Europe and the United States, no one imagined the cataclysm that lay ahead.

A major reason was the perceived irrationality of great power conflicts. This was the theme of two major works, Jan Gotlib Bloch's *Is War Now Impossible?* (1898) and Norman Angell's *The Great Illusion* (1909). The latter was a huge bestseller that was immediately translated into 11 languages; it would give its name to Jean Renoir's classic antiwar film, *Grand Illusion*. The "Great Illusion," British journalist Angell argued, was that nations gained by armed confrontation, militarism, war, or conquest. "War, even when victorious, can no longer achieve those aims for which people strive." Moreover, "The forces which have brought about the economic futility of military power have also rendered it futile as a means of enforcing a nation's moral ideals or imposing social institutions upon a conquered

---

35 See Platt, Stephen R. *Autumn in the Heavenly Kingdom: China, the West, and the Epic Story of the Taiping Civil War*. Atlantic Books, 2014.

people."[36] Bloch, a Polish financier-turned-student of war, portrayed the futility of any war once started. Technology had rendered maneuvers over open ground obsolete. A major war would be one of entrenchment and stalemate with no clear victor. The result would be severe economic and social dislocations involving "the breakup of the whole social organization" and revolutions from below.

Their talk of war's "impossibility" has led some to dismiss Angell and Bloch as poor visionaries. Just a few years later, 1914's guns of August began World War I putting the world on the path to the violence of the 20th century. All their claims about the irrationality of war were vindicated as the Great War only brought social disruption and ruin.

But rather than leading to the renunciation of war, it laid the groundwork for an even greater war to follow.

[2]/ The Violent 20th Century

Historians wrestle to explain why the 20th century worked out as it did. But how to explain the violence of the 20th century is part of the larger question of how to tell the story of the 20th century generally.

A standard view sees it as a morality tale, one of triumph of good over evil. Ferguson summarizes it thus: the story of the 20th century is

a kind of protracted, painful but ultimately pleasing triumph of the West. The heroes (Western democracies) were confronted by a succession of villains (the Germans, the Japanese, the Russians) but ultimately good always triumphed over evil. The world wars and the Cold War were thus morality plays on a global stage.[37]

This narrative both reflects and supports the celebration of World War II as the archetypal "good war." It is the conflict that confirms—against any pacifist doubts—the necessity of war to triumph over evil; hence, the need for ongoing military expenditures should such a challenge again arise.

This is the narrative that was fashioned around mid-century. It was taught to Baby Boomers like me—who, not surprisingly, were taught almost nothing about World War I. For the power of this narrative rests on *forgetting* that war, which George Kennan termed the 20th century's "great seminal catastrophe." The rash of books that began appearing in the 1990s surveying the 20th century all confirmed this. There wouldn't have

---

36 Angell, Norman. *The Great Illusion: A Study of the Relation of Military Power to National Advantage.* 4th ed., CreateSpace Independent Publishing Platform, 2015, pp. 4–5.
37 Ferguson, Niall. *The War of the World: Twentieth Century Conflict and the Descent of the West.* Penguin Press, 2006, p. xxxvii.

been any World War II (the "Good War") without World War I (which its soldiers called "The Big Fuck-Up"). Without World War I, there would have been no Bolshevism, no Great Depression, and no Fascism. Indeed, the links are so intimate that the two world wars are now regarded as *one* war, akin to the 17th century's Thirty Years War. And the story did not end there, if one regards World War II as bleeding into another 30-year "Cold War," with the century's other great war-induced revolution in China.

From this perspective, the tale of the 20th century looks very different. It is not a story of war (World War II) as the *solution* to evil. It is a story of war (World War I) as the *source* of evil.

So, the question posed by the 20th century is:

How do we explain World War I in the first place?

[3]/ Imperial Fantasies

This is where the pacifist perspective departs from others. Accounts of World War I seek to identify some deeper logic in why it occurred, hence some larger logic to the century that it defined. But what pacifists see is an *absurdity* that casts its shadow over the entire age.

The standard perception of World War I is deeply informed by a fixation on the Western Front which sees the war as a straightforward conflict between European states. In fact, the war was a war between empires around imperial ambitions. All the contending parties in 1914 were empires, either traditional dynastic empires or nation-state–empires. The British and French immediately attacked and occupied German colonies in Africa and the Pacific; Russia hoped to extend its influence at the expense of the Ottoman Empire; Turkey's actions were aimed at preventing further erosion of that empire; Italy joined on the Allied side in the hope of imperial territory in the Balkans and the Mediterranean; Japan's entrance into the war involved imposing concessions on China that laid the groundwork for World War II's Pacific conflict. Germany's imperial ambitions were most fully realized in the Treaty of Brest-Litovsk where the Bolshevik government ceded the whole of the western Russian Empire, a landmass greater than anything Hitler would occupy 25 years later.

Other than the power jockeying of empires, though, no one could really say what the conflict was *about*. Thomas Mann observed that the war was "unpolitical," meaning it was detached from any concrete political/economic realities. John Keegan states more broadly "Politics played no part in the conduct of the First World War worth mentioning."[38] Astonishingly, the principal demand of antiwar forces was that the contending sides

---

38 Keegan, John, *Op. cit.*, p. 21.

simply state what they were fighting *for*. It was only in the war's very last phase, in response to the Bolsheviks' publishing secret agreements between the Allies, that any significant efforts were made to explain what the fighting was all about.

And what drove all this power jockeying was a set of delusions which historian Birthe Kundrus has termed "imperial fantasies." They were rooted in the century-long experience of European empire, and would provide the ideological context for World War II as well. They were fantasies about "the settlement of wild frontiers, or the prospect of an Eldorado of riches, or an exalted 'civilizing mission', or the fulfillment of a manifest destiny that would reinvigorate the nation."[39] At the heart of the 20th century's delusions were the notions of "race" and "space." Pseudo-scientific Social Darwinist thinking portrayed empires as the champions of something called "races" that were conceived as in a life-and-death struggle over "racial" dominance versus "racial" subjugation. Another 19th-century legacy was the notion of "living space," whereby "superior" cultures were conceived as not having enough "breathing room" to expand according to their destiny. Hence, the only solution to this political/economic claustrophobia was stealing the territory of others.

But here is what pacifists see when they look at World War I.

They see a war to "make the world safe for democracy" that led to the most brutal forms of totalitarianism. They see a war to "defend civilization" that dragged it into deeper forms of barbarism. Above all else, they see a "war to end all war"—that led to a century of unprecedented warfare.

This was its true innovation in the ideology of warfare. Wars had previously been portrayed as defenses of "civilization" and its political ideals. But it's hard to find a precedent for the idea of fighting a "war to end war," that is, not just a war to end a particular war and bring peace but a war to end *all* war and thus secure peace *forever*. Yet it was a natural outgrowth of modern, total war in this sense: an enterprise so destructive and absurd could only be justified by such an ultimate end. Defending "tiny Belgium" from German predations could not fit the bill. So, as the full horror of World War I became apparent, the best argument for it was that it would ensure that this sort of thing never happened again. Ever since, a central theme of war has been the ending all war through war.

---

39 Kundrus, Birthe. "Die Kolonien—'Kinder des Gefühls und der Phantasie'." *Phantasiereiche*. Edited by idem, pp. 7–18 [Cited in Overy, Richard. *Blood and Ruins*. Allen Lane, an imprint of Penguin Books, 2021, p. 880].

Reflecting on World War I, John Keegan speaks of it as "an extraordinary, a monstrous cultural aberration."[40] But can the defining event of the 20th century really be so easily dismissed? Pacifists held otherwise. World War I was no aberration, but the inevitable result of a "war system" that had arisen with the modern European state and would continue to generate war until the war system's grand illusions were rejected. But this was no easy task. It had taken three centuries to construct the "war system." It would take a long time to deconstruct it.

World War I ended when people got sick of fighting it. Many had expected the war to go on much longer, but they did not anticipate the revulsion against the whole enterprise. The first major break was Russia's defection following the Bolshevik revolution. This resulted directly from disenchantment with the conflict, especially among deserting soldiers that the Bolsheviks mobilized via their demand for "Peace, Bread, and Land." Plus there were outright mutinies by Allied soldiers on the Western Front that scholars regard as impacting Allied strategies. The most famous was the 1917 French mutiny involving almost half of all French infantry divisions. The most significant mutiny among the British was the 1917 Étaples mutiny at a training camp in France involving thousands of British and British Imperial soldiers. The German government sought an armistice when its own military collapsed in conjunction with the specter of a war-induced Bolshevik-type revolution back home. Many concluded that the problem of war could not be addressed one war at a time; it required a more systematic solution. The international arrangements like the League of Nations reflected the heartfelt attempt by many to prevent such a cataclysm from happening again. Many pacifists supported these efforts.

The principle proposed as the basis of the new order, "national self-determination," appealed to the aspirations of non-European peoples. Unfortunately, it laid the basis for the next major conflict by valorizing another grand illusion of the 20th century: *nationalism*. It has continued to stoke the fires of war to this day.

[4]/ The (One) "Good" War

I shall speak to World War II at several points in this work. At least for my generation, it has loomed over any discussion of war's legitimacy and has been taken as a decisive counterargument to pacifism. "Yes, war has been almost entirely a story of futility, stupidity, and waste—but there has been at least *one* good war: World War Two! And what does the pacifist say about *that*?"

---

40 Keegan, *Op. cit.*, p. 21.

I take World War II seriously. My father served in that war, though he came away from his war experience as a committed pacifist. I take any assessment of that war seriously, meaning it requires serious attention to the event itself—more attention, I think, than it receives when people simply throw the example out as a way of ending the discussion. Much of the recent scholarship on World War II has dwelled on the sheer amount of obfuscation and mystification that has surrounded it, due in no small part I think to its ideological importance in the argument for war.

Here are some concerns to be noted at the start:

- The first issue confronting any appeal to World War II is what is *meant* by World War II.

There are two issues here. One has been anticipated, of whether it is legitimate to separate World War II from World War I for purposes of appraisal, or whether they are rightly regarded as a single conflict. Richard Overy speaks from a consensus when he writes, "little is to be gained by separating the two giant conflicts. Both can be seen as stages of a second Thirty Years War about the reordering of the world system in a final stage of imperial crisis."[41] It is like someone who wants to hive off one phase of the 17th-century religious wars from the rest and construct a case for war generally on its justice.

The other issue is the Eurocentrism of ignoring the war's impact on the rest of the world. As with World War I, when people talk about World War II they have in mind the conflict in Europe, and specifically the conflict in Western Europe between Germany and the American/British—with the specter of Hitler seeking world domination. This is not in fact how most Americans thought of it at the time. They regarded the Pacific War as more important, whose stakes were not about world domination but straightforward self-defense. More to the point, though, the impact of the war outside of Europe is entirely ignored. Consider, for example, the 1943 Bengali famine: scholars regard it as a direct result of Allied wartime priorities, specifically sending food elsewhere for military reasons. (Churchill blamed the famine on the fact that "Indians were breeding like rabbits.") As a result, an estimated two million to four million people perished; even by the lower figure, more Indians died in this famine than died as combatants in the two world wars, in the independent struggle, and in the genocide of partition combined.[42]

---

41 Overy, Richard J., *Op. cit.*, p. xii.
42 See Hastings, Max. *All Hell Loose: The World at War 1939–1945*. Harper, 2011, pp. 422–426; Weinberg, Gerhard. *A World at Arms—A Global History of World War II*. Cambridge University Press, 1994, p. 493ff.; Mukerjee, Madhusree. *Churchill's Secret War: The British Empire and the Ravaging of India during World War II*. Penguin Books, 2018.

Thus, the "good war."

- A second issue is that what makes World War II "good" in many people's eyes has little to do with what the war was actually *about*.

The title of Richard Overy's recent study of the conflict is *Blood and Ruins: The Last Imperial War, 1931–1945*. Michael Mann sees it as a war between the established imperialisms of Britain, the United States, and the Soviet Union, on the one hand, wanting to keep international arrangements in place, and the upstart empires of Germany, Italy, and Japan, wondering why they could not have their "place in the sun" too. Viewed thus, Mann regards World War II as "the culmination and hubris of European traditions of militarism and imperialism, now exported to others."[43] One could argue that very good things were nevertheless achieved by the war. But any such argument would have to weigh them against the very bad things that resulted from it as well. Whether or not it was the "last" imperial war or just another imperial war, leading to a new phase of empire, remains to be considered.

- A final issue to be noted for now is the force of singling out one "good" war as a case for war *per se*.

There have been approximately 14,600 wars in recorded history, approximately 3,200 major ones. One might reasonably ask if the fact that we can agree on just *one* case of a "good war" counts in favor of war—or *against* it. Imagine a hypothetical social practice, call it "Brunting," that has been around for millennia, has generated great physical suffering, and has required massive social organization to pursue it. Finally, suppose that in all the history of "Brunting" people could agree on only *one* instance in which "Brunting" made any sense at all, while all the others were seen as the height of insanity. Would anyone take this single case as a decisive argument in *favor* of "Brunting" and as refuting—in one fell swoop—those (pacifist-type) "Anti-Brunters" who urged that the whole sordid practice be relegated to the dustbin?

Ultimately, World War II was portrayed as another "war to end war." The Atlantic Charter, the closest Britain and the United States came to stating their war aims, concluded with the "belief" that "all of the nations of the world, for realistic as well as spiritual reasons, must come to the abandonment of the use of force." The United Nations Charter began with

---

43 Mann, Michael. *The Sources of Social Power: Vol. 3*. Cambridge University Press, 2012, p. 423.

the commitment "to save succeeding generations from the scourge of war, which twice in our life-time has brought untold sorrow to mankind." As with World War I, the desire for peace was a motivating fact for soldiers. My father enlisted in the war because he took the Atlantic Charter seriously; he never, for a moment, thought that the United States would be subjected to foreign conquest. He, like soldiers throughout the 20th century, went to war to end war. Pacifists applauded the ideals of global collaboration embodied in the United Nations. They took seriously President Franklin Roosevelt's "one-world" vision.

So the question is why this war to end all war led yet again to *more* war. Part of the answer lies in the political arrangements that emerged from it, involving the triumphant political structure: the *nation-state*.

At the conclusion of World War I, the principle of national self-determination meant that borders were invented and adjusted to fit national groupings, whose populations were generally left in place. The problem was that this made absolutely no sense for Europe as it was historically constituted. "The continent of Europe was once an intricate, interwoven tapestry of overlapping languages, religions, communities and nations," writes Tony Judt. "Many of its cities—particularly the smaller ones at the intersection of old and new imperial boundaries, such as Trieste, Sarajevo, Salonika, Cernovitz, Odessa or Vilna—were truly multicultural societies *avant le mot*."[44] There was no coherent way to impose state structures on it that would fit nations to states. So, after 1945, the approach was the opposite: borders stayed largely intact, and peoples were moved to fit the borders. Hitler and Stalin had already pursued population transfers. These continued after the war, from the belief that political stability required the homogeneity of populations. Churchill told the British House of Commons in 1944 that deportations would provide the "most satisfactory and lasting" solution to ethnic problems. The upshot was that World War II and its aftermath saw the greatest instances of forced migration in human history. At the time, some protested the fate of Germans, about 12 million of whom—most of them women and children—were driven back to the new German borders. But Poland exemplified the creation of this new "monotone" Europe. Prior to World War II, ethnic Poles had constituted barely two-thirds of its population; after the war, shorn of its Jewish, Ukrainian, and German minorities, they constituted 98% of the population. Hence, in Judt's words, the upshot of the two world wars was that "multicultural" Europe was "smashed into dust." "The tidier Europe that emerged, blinking, into the second half of the twentieth century had fewer loose ends. Thanks to war, occupation, boundary adjustments, expulsions

---

44 Judt, Tony. *Postwar: A History of Europe since 1945.* Penguin Press, 2005, p. 8.

and genocide, almost everybody now lived in their own country, among their own people."[45] Whatever possibilities for conflict this involved were muted by the division of Europe by the Cold War, which we turn to shortly.

Outside Europe, though, the imposition of the nation-state model only led to instability, as borders that were arbitrary artifacts of the European empire were frozen in places where they had never made any sense. The instability that had defined interwar Europe was now exported into post-war everywhere else.

[5]/ The Cold War and Martial Liberalism

The end of World War II ushered in the end of the colonialist phase of European imperialism, a dominant political form for 500 years. As in the late 19th/early 20th centuries, a focus on the West saw it as a time of peace. John Lewis Gaddis has termed the Cold War a "long peace," meaning an absence of direct great power conflict. In fact, colonialism's end was not a peaceful one but inaugurated a new phase of war. The response of co-lonial powers to local insurrections was a violent one; recent scholarship has stressed the violence of the British in this regard.[46] And the internal conflicts it involved were significantly exacerbated by superpower conflict. Both the United States and the Soviet Union were anti-colonialist in their official ideologies, so both sought to exploit such conflicts in ways that would further their ambitions. Those ambitions were driven by the tra-ditional paranoias of empire: the United States seeking to "contain" the Soviet Union, the Soviet Union resisting being "encircled" by the United States. Hence, both poured human and material resources into otherwise local conflicts in ways that vastly magnified the importance that such con-flicts otherwise possessed. (Eighty cents out of every dollar in American and Soviet foreign aid to the Third World was devoted to military conflict.)

As always, it is a mistake to assess the wars of this era one by one. The Cold War, as one of its leading scholars insists, must be understood as a system, "a *network of connections* that linked the struggles together and increased their destructive potential by an order of magnitude."[47] That destruction was principally inflicted on the innocent. The Cold War con-tinued the war system's long-term trend to killing civilians over soldiers.

The struggles proceeded in three phases, each connected to the demise of previous colonialisms. The collapse of the Japanese empire led quickly

---

45 Judt, Tony, *Op. cit.*, p. 9.
46 Elkins, Caroline. *Legacy of Violence: A History of the British Empire. Vintage*, 2023.
47 Chamberlin, Paul Thomas, *Op. cit.*, p. 4.

after World War II to revolutions in East Asia, principally the Chinese Revolution but also Korea, a traditional Japanese colony. The global Communist revolution was now seen as a real contender for world power, inaugurating the first conflicts of the Cold War. American actions in Korea were especially noteworthy for their violence. The United States dropped significantly more bombs in Korea (not counting over 30,000 tons of napalm) than in the entire Pacific theater in World War II. The second phase involved the collapse of European colonialisms, most significantly in Vietnam where the United States stepped in to replace France. With the emerging tensions between Moscow and Beijing, Indo-Asia became the site of triangular struggles between these powers and America that dramatically magnified the bloodbaths in Vietnam, Indonesia, Bangladesh, and Cambodia. But the conflicts between the major communist powers meant that the specter of progressive revolution there and in the developing world generally was diminished and eventually extinguished. By the late 1970s, global communism had collapsed but the era of violence just assumed new forms. The third phase of conflict saw it continuing to move westward toward the Middle East and the challenges to the colonialist empires and their progeny. The conflicts now were framed in terms of religious sectarianism and ethnic identity, the focus of the Cold War's final decade. By the mid-1980s, America's Central Intelligence Agency had become the largest patron of rebel armies in the world, pumping millions of dollars into conflicts in Afghanistan and elsewhere.

If America's actions in the Cold War finished communism as a political force, their principal impact on future conflicts was to cripple secular liberation movements throughout the Third World so as to destroy the principal alternative to the sectarian/ethnic politics that now confronted America with a vengeance.

What I have termed the Paradox of European Modernity has been the intimate relation between war and the growth of political and social rights due to the bargains that rulers have made to enlist support for their war projects. Nowhere has this been truer than in the United States where, for example, every extension of the voting franchise has been the result of war (most recently granting the 18-year-old vote in return for Vietnam War service). My term for this is the *war contract* at the heart of the war system. Its bargain—of political and economic rights to citizens in return for their service/support for war efforts—achieved full fruition in the "warfare–welfare states" of the World War II–post-World War II era. Indeed, the American strategy for confronting the Soviet Union involved extending the warfare–welfare model to its global allies in the support of both military and economic development.

The ideological form this took in the United States is what I have termed *martial liberalism*. Liberalism had traditionally been deeply skeptical of war, whose collective impulses seemed directly contrary to the claims of the individual. So too socialism, whose antiwar dimension had been paramount prior to World War I. One upshot of the 20th century and of World War II especially was the *militarization* of these ideologies. Liberalism's fate was the stranger of the two in its accommodation with the claims of the all-powerful all-intrusive national security state. It was the ideology of the hysterical "moderate": one who trumpeted liberal values as exemplifying "reasonableness" over rashness, while insisting that the defense of those values could warrant subjecting the world to nuclear holocaust. Sociologist C. Wright Mills spoke of how the irenic impulses of liberalism had been lost in what he called its "military metaphysics."[48]

It is hard for younger people to grasp how pervasive martial liberalism's "military metaphysics" was for someone of my generation. It wasn't just that conscription was accepted without question; indeed, prior to the Vietnam War, conscription consistently ranked among the most popular American institutions! It infused our education. Everyone has heard of how we crouched beneath our desks to shield ourselves from nuclear explosions (in Los Angeles, this happened every Friday). Only some of us may remember President Kennedy's "President's Fitness Program," driven by the concern that America's youth were getting too soft for what Kennedy termed the "long twilight struggle against communism." Once a month we were required to go out and run a one-hundred-yard dash. It was never clear to me if this was training me to run *toward* the "communists" or *away* from them.

[6]/ The Grand Disillusionment

In time, this era proved to be the twilight of martial liberalism. Sir Michael Howard has described the disenchantment with "nationalized war" due to the violent excesses of the two world wars, their boundless demands on the ordinary citizen, and the clandestine machinations and hypocrisies in the Cold War.[49] It proceeded differently in different countries, as James J. Sheehan recounts in his study of postwar disillusionment *Where Have All the Soldiers Gone?: The Transformation of Modern Europe*. Its demise in the United States is commonly ascribed to the Vietnam War, but the matter proved complicated. Martial liberalism maintained a ghostly existence in the Reagan era when saber-rattling against the Soviet Union rationalized a massive upsurge in military spending. But all this military posturing was detached from any thought of personal sacrifice of the sort that had

48 Mills, C. Wright. *The Causes of World War Three*. Greenwood Press, 1976.
49 Howard, Michael. "War and the Nation–State." *Daedalus*, 108(4), 1979, pp. 106–107.

previously marked war efforts, like military service or higher taxes. It was the dawn of what I have termed *alienated war*, the basics of which persist in the United States to this day.

In the end, what most contributed to Americans' disenchantment with war were two social movements: the civil rights movement and feminism. By challenging racism, the civil rights movement challenged a central feature of American wars against foreign peoples, as well as the genocidal wars against native peoples by which Euro-Americans had constituted the nation itself. Hence, the leading antiwar voice of the 1960s, Dr. Martin Luther King, Jr., was also its leading civil rights advocate. By challenging sexism, the feminist movement challenged the gendered ideology of manliness that had long been a pillar of pro-war thinking. Hence, the leading voices in the anti-nuclear movements of the late 1970s and 1980s were women such as Dr. Helen Caldicott and Randall Forsberg.

The Cold War ended with the collapse of militarized socialism in the Soviet Union. But while it was trumpeted as a victory for America's martial efforts, it was not received as an endorsement for war generally. On the contrary, it engendered a brief bloom of optimism that humanity might finally recognize the ultimate pointlessness of armed conflict. This coincided with the larger recognition among scholars that the century just then concluded had been a century of war.

Some were impressed by the apparent disappearance of war between major powers. Evan Luard wrote,

> Most startling of all has been the change that has come about in Europe, where there has been a virtual cessation of international warfare . . . . Given the scale and frequency of war during the preceding centuries in Europe, this is a change of spectacular proportions: perhaps the single most striking discontinuity that the history of warfare has anywhere provided.[50]

Sir Michael Howard agreed, "It is quite possible that war in the sense of major, organized armed conflict between highly developed societies may not recur."[51] John Keegan wondered if this disillusionment with war did not extend to the world generally. "Despite confusion and uncertainty, it seems just possible to glimpse the emerging outline of a world without war."[52] John Mueller stressed the role of antiwar movements. The West had experienced destructive wars before which generated peace initiatives

---

50 Luard, Evan. 1986. *War in International Society*. Yale University Press, 1986, p. 77.
51 Howard, Michael. *The Lessons of History*. Yale University Press, 1991, p. 176.
52 Keegan, *Op. cit.*, p. 51.

in response, but nothing like the late 20th century's disillusionment. The new factor was a strong and articulate antiwar movement dating from the 19th century and increasingly prominent in the 20th century. In Mueller's words, the conflicts of the 20th century seemed to be "a colossal confirmation of [the anti-war movement's] gadfly arguments about the repulsiveness, immorality, and futility of war and of its uncivilized nature."[53]

[7]/ After 9/11: "Its Going to Be Long"

This optimism ended abruptly with the events of 9/11, just as August 1914 had shattered an earlier era of hope.

And, like August 1914, 9/11 immediately provoked a pro-war hysteria suggesting that such bellicose sentiments had never gone away. Some spoke breathlessly of how it was nothing less than the start of "World War IV" (World War III having been the Cold War). Former head of the CIA R.J. Woolsey wrote,

> We witnessed three global wars in the past century. Only a decade ago we ended WW III, also known as the Cold War. But a new enemy has been on the march and we have entered a dangerous but subtler conflict: World War IV.

The war would be fought "heavily on the home front and so we must improve the resilience of our infrastructure to assure our capacity to win a long, difficult conflict."[54] The enemy, Norman Podhoretz warned, would be "even more dangerous and difficult to beat" than Nazi Germany or the Soviet Union.[55] Some saw it as a moment of great opportunity in which America, having failed to impress its "Western values" on the world by force of example, could now do it by force of arms. Secretary of Defense Rumsfeld maintained that the United States now had the kind of opportunities that World War II offered to refashion the world. "People are now coming out of the closet on the word 'empire,'" Charles Krauthammer wrote.[56] All of which prompted left-wing scholars like Gabriel Kolko to fear that it presaged "another century of war."[57]

53 Mueller, John. "War Has Almost Ceased to Exist: An Assessment." *Political Science Quarterly*, 124(2), 2009, pp. 297–321.
54 Woolsey, R.J. *WWIV: Who We're Fighting- And Why.* Vol. 4. Rich. J. Global L. & Bus. 1, 2004.
55 Podhoretz, Norman. *World War IV: The Long Struggle against Islamofascism.* Doubleday, 2007, p. 230.
56 Eakin, Emily. "It Takes an Empire." *New York Times*, April 2, 2002.
57 Kolko, Gabriel. *Another Century of War?* New Press, 2004.

Voices across the political spectrum agreed that America's post-9/11 invasion of Afghanistan starting its "Global War on Terror" was an exemplary "just war"—which only unreconstructed peaceniks could oppose. Longtime peace advocate Richard Falk wrote that this was "the first truly just war since World War II."[58] A *Wall Street Journal* editorial by Scott Simon was titled "Even Pacifists Must Support This War."[59] *The American Prospect*'s Robert Kuttner now proclaimed that only people on the "extreme left" could believe this was not a just war.[60]

Not surprisingly, doubts about the whole enterprise quickly arose from more sober quarters, and not just those on the extreme left. In an eerily prescient *Foreign Affairs* article, "What's in A Name? How to Fight Terrorism," Sir Michael Howard warned

[T]o use, or rather to misuse, the term "war" is not simply a matter of legality or pedantic semantics. It has deeper and more dangerous consequences. To declare that one is at war is immediately to create a war psychosis that may be totally counterproductive for the objective being sought.

He went on to ask if the "War on Terror" should be properly termed a "war" at all, given the open-ended ambiguous nature of the enterprise.[61] Replies to Sir Michael's warning insisted that yes, it was a war—but not of the "traditional" type. The case against calling the "War on Terror" a "war," wrote one authority, rested on the mistaken assumption that wars must have a beginning, a middle, and an end, that their aims must be clearly stated, or stated at all, that they must be fought by recognizable combatants, and must lead to one side or another winning. But today we were dealing with what another authority called "non-linear war," to which none of these notions applied; indeed, the whole distinction between war and peace was blurred.[62] Other names proposed were "hybrid wars,"

58 Falk, Richard. "Defining a Just War Ends and Means." *Nation Magazine*, October 11, 2001.

59 Simon, Scott. "Even Pacifists Must Support This War." *Wall Street Journal*, October 11, 2001.

60 Cited in Zinn, Howard. "A Just Cause, Not a Just War." *The Progressive*, December 2001.

61 Howard, Michael. "What's in a Name? How to Fight Terrorism." *Foreign Affairs*, 81(1), 2002, pp. 8–13.

62 Gerasimov, Valery. "The Value of Science in Prediction." *Military-Industrial Kurier*, 2013, http://inmoscowsshadows.wordpress.com/2014/07/06/the-Gerasimov-doctrine-and-Russian-non-linear-war/.

"postmodern wars," and "wars of the third kind."[63] If we just scuttled such archaic notions, if we conceived of war instead as something with no defined beginning or end, no particular aims, no clear adversaries, and no decisive outcome—then enterprises like the "War on Terror" fit right in! Some in the Pentagon adopted the term "the Long War" for a conflict expected to last 50 years or more. In his 2005 *Winning the Long War: Lessons from the Cold War for Defeating Terrorism and Preserving Freedom*, James Jay Carafano of the Heritage Foundation wrote, "We can't agree it's global, we can't agree it's terrorism, but we all generally agree it's a war . . . [and] it's going to be long!"[64]

Plus there were historical precedents for such a war, some scholars claimed. One authority found them in ancient times, likening them— totally without irony, as it ended rather badly—to "the kind of long struggle with exterior barbarians that characterized the wars of the later Roman Empire."[65] Others likened it to more recent experiences of colonialism. Philip Bobbitt concluded *The Shield of Achilles* by likening America's predicament to that of white settlers circling the wagons against the onslaught of Native American "savages." The events of 9/11 were "the herald of further savagery and the call for defenses," for a war that will have "no final victory," just the ongoing project of "avoiding defeat."[66]

Can anyone imagine a more compelling case *against* a war than this picture of it as an endless, ambiguous enterprise that likens us to the residents of Rome in its final days or early Euro-American settlers besieged by "savages"?

If some think they have encountered this picture of war before, they have—in George Orwell's 1984. "War has changed its character," Orwell wrote in that dystopian classic. The fighting "takes place on vague frontiers whose whereabouts the average man can only guess at." Before, war was something that "sooner or later came to an end, usually in unmistakable victory or defeat," but now it is "literally continuous" to the point that the whole

---

63 Duffeld, Mark. *Global Governance and the New Wars: The Merging of Development and Security*. Zed Books, 2001; Gray, Chris Hables. *Post-Modern War: The New Politics of Conflict*. Routledge, 2007; Hoffman, Frank. *Conflict in the 21st century: The Rise of Hybrid Wars*. Potomac Institute for Policy Studies, 2007; Kaldor, Mary. *New and Old Wars: Organized Violence in a Global Era*. Polity Press, 1999; Munkler, Herfried. *The New Wars*. Polity Press, 2005; Smith, Rupert. *The Utility of Force: The Art of War in the Modern World*. Alfred A. Knopf, 2005.

64 Cited in Graham, Bradley, and White, Josh. "Abizaid Credited With Popularizing the Term 'Long War'." *The Washington Post*, February 3, 2006.

65 Brown, C. "Reflections on the 'War on Terror', Two Years on." *International Politics*, 41, 2004, pp. 51–64.

66 Bobbitt, Philip. "Postscript—The Indian Summer." *The Shield of Achilles*, reprint edition. Anchor Books, 2009, pp. 819–824.

distinction between war and peace "has ceased to exist." "Strictly speaking, it has not always been the same war," though "to trace out who was fighting whom at any given moment would be literally impossible." In contrast to the mass wars of the past, war now involves "very small numbers of people, mostly highly trained specialists, and causes comparatively few casualties." But this does not mean that attitudes are "less bloodthirsty or more chivalrous." On the contrary, "war hysteria is continuous and universal." Hence, the enemy, whoever they are at the time, "always represent absolute evil, and it follows that any past or future agreement with them is impossible."[67]

The "War on Terror"/"World War Four" chiefly differed from previous ones in its alienated character. Immediately after 9/11 military enlistments plummeted. The Iraq War would be crippled by a dearth of soldiers, despite desperate attempts by the military to lower all standards for enlistment and to prohibit those who had enlisted from leaving. Of special note was the resistance of African Americans and Latino Americans from serving, a principal source of American soldiers. The massive cost of the "War on Terror" would not be met by raising taxes as in previous wars. Rather, taxes were cut and citizens were encouraged to go shopping in the name of the war effort. The amorphousness of the whole enterprise was obscured by the hyper-moralization that infused popular discourse. Those killed, whoever they were and wherever they happened to be, were "bad guys," and the soldiers doing it were constantly "thanked for their service" by citizens who had no idea what their "service" involved or why (or where) it was being performed.

The pro-war hysteria did not last. One upshot of alienated war's detachment of the ordinary citizen from the war effort is that passions around it are ephemeral. The post-9/11 pro-war bravado proved to be a "sugar high" that would disappear as quickly as it came. It was just a matter of time until the ambiguities of the whole enterprise came back to haunt it. The term "forever war," like "Long War," had been first proposed both to describe and to endorse the kind of challenge that the "Global War on Terror" posed. Now it became a term of abuse, with politicians promising to end them. Former Army officer Douglas Lute, who had overseen operations in Afghanistan for both the Bush and Obama administrations, admitted of the war in Afghanistan "We didn't have the foggiest notion of what we were undertaking."[68]

67 Orwell, George. *1984*. Secker & Warburg, 1949.
68 Quoted in Whitlock, Craig. "At War with the Truth." *Washington Post*, December 9, 2019.

The wars following 9/11 killed more than 900,000 people, displaced 38 million people, and cost United States an estimated $8 trillion. What lessons were drawn from all this? Orthodox opinion responded with much the same evasions that followed the Vietnam War. The "Global War on Terror" had been mishandled, its legitimate aspirations betrayed, its disasters all a mistake. The exception was a spate of studies that appeared wondering why, starting with Korea, the United States seemed incapable of decisively winning *any* wars—despite being the greatest military power in the history of the earth.[69] Various explanations were offered, but the principal one was that the country was insufficiently clear on why it fought wars in the first place. Donald Stoker wrote that the US and other modern liberal democracies "too often fail to clearly define what they're fighting for."[70]

So the pacifist challenge to World War I—What, precisely, are we fighting for?—pertained to all of America's endeavors.

[8]/ Cold War Redux

Once again, the self-reflection did not last long. Even before the "Long War" petered out, official opinion was pivoting from the enemy as "global terrorism" to the insistence that the world was back in a new "Cold War," except now it was a three-way conflict between the United States, China, and Russia. (A cynic might ask if America would have been better off remaining in the *last* Cold War.)

What, precisely, was this new global conflict *about*?

Official discourse avoided any specific talk of aims by reverting to a "great power" rhetoric of the type that characterized 19th-century crisis diplomacy and laid the basis for World War I. Official opinion saw "adversaries" everywhere. American voices were especially alarmed about China and its perceived quest for "hegemony" in Asia. But the main worry was the Social-Darwinist-type concern that the strength of one country was inherently a threat to others. Former Department of Defense official Elbridge Colby wrote in his *The Strategy of Denial: American Defense in an Age of Great Power Conflict* that Chinese "dominance" in Asia

---

69 Books on this topic include Stoker, Donald. Why *America Loses Wars: Limited War and US Strategy from the Korean War to the Present.* Cambridge University Press, 2022; Ullman, Harlan. *Anatomy of Failure: Why America Loses Every War It Starts.* Naval Institute Press, 2017; Tierney, Dominic *The Right Way to Lose a War: America in an Age of Unwinnable Conflicts.* Little Brown and Company, 2015.

70 Stoker, *Op. cit.*, p. 5.

would "compromise Americans freedom, prosperity, and even physical security."[71] The "Long War" was being replaced by the "Long Game," eerily reminiscent of the 19th century's "Great Game" of endless conflict in India/the Middle East between that era's "great powers." Rush Doshi, a Biden administration official, wrote in *The Long Game: China's Grand Strategy to Displace American Order* that America and China were now locked in a life-and-death struggle over regional and global order. Unless it could check Chinese aspirations, the United States was in danger of losing its "preeminent position" worldwide.[72]

As I write this, Russia is engaged in a criminal invasion of a country directly on its border, Ukraine. As usual, the logic of the war system requires that any response to it must ignore the hypocrisies surrounding it. Two decades ago, the United States conducted a massive invasion of a country half a world away, with the full support of Ukraine—which, after the United Kingdom, was America's greatest ally in the effort. Now we are told by the former cheerleaders of that effort that Russia's invasion of Ukraine is an "unprecedented" violation of the global legal order that the "West" has sought to maintain, and is the latest threat to "civilization" as we know it. Whatever else Russia might be, it is *not* a military threat to the "West." The 2021 Russian military budget was $66 billion/year. The American military budget was $801 billion/year, and the combined NATO military budget was $1,164 billion/year. Hence, the Russian military budget is about 8% that of America's and about 5% that of NATO's. (Russian GDP is about $1.5 trillion a year, smaller than Canada's, significantly smaller than Brazil's, and about 7% that of the United States.) Since the war began, Russia has proved itself unable to conquer the eastern parts of Ukraine constituting about 6,500 square miles, or an area slightly bigger than greater New York City.

Yet "the West" is gearing up once again to face the Russian "threat."

There is in fact reason for alarm. The United States' 2018 National Defense Strategy began by stressing that "Inter-state strategic competition, not terrorism" was now the "major concern." It proceeded with the usual alarm that the American military had deteriorated on every level, due to insufficient funding. It proceeded to list the five major challenges to America's "security interests," with priority given to China and the emphasis on competing with it in the South China Sea. It concluded with the need to make the United States forces more "lethal, agile, and resilient," all in line with facilitating the emphasis on "fighting and winning conflicts with

71 Elbridge, Colby. *The Strategy of Denial: American Defense in an Age of Great Power Conflict.* Yale University Press, 2021, p. 10.
72 Doshi, Rush. *The Long Game: China's Grand Strategy to Displace American Order.* Kalorama, 2022.

China or Russia." Really—"*winning*" conflicts with Russia and China? Even at the height of the Cold War, American leaders were reluctant to speak of "winning" a military conflict with the Soviet Union, but now that is official military policy.

As I write this, General Mark Milley is the chairman of the United States military's Joint Chiefs of Staff, having previously served as the Army's Chief of Staff. In his current post, General Milley is well positioned to represent official military thinking. An article titled "General Milley Predicts Grim Future of Deadly Great Power Wars Fought in Cities" reported his 2022 speech to the United States Military Academy's graduating class.[73] It reported, "America's highest-ranking military officer painted a picture of a dark future with great power wars fought in urban environments." Speaking to graduating cadets at the United States Military Academy, General Milley forewarned of death tolls for US soldiers in the tens of thousands. "We are entering a world that is becoming more unstable. The world you are being commissioned into has the potential for significant international conflict between great powers, and that potential is increasing, not decreasing." After listing America's new arsenal of robotic tanks, invisible airplanes, and other technical doo-dads, General Milley concluded, "the coming wars will exact high tolls on civilian populations."

"Fighting in cities" with "high civilian tolls"—between "the great powers": we seem to have come full circle to the pre-World War I era.

Contemporary critics of this are returning to the writings of Cold War critic C. Wright Mills. Mills coined the phrase "crackpot realism" to describe thinking about war that preached absurd conclusions while presenting them as the height of rationality. In *The Causes of World War Three*, Mills characterized the mindset of such military-focused policymakers thus:

> They know of no solutions to the paradoxes of the Middle East and Europe, the Far East and Africa except the landing of Marines. Being baffled, and also being very tired of being baffled, they have come to believe that there is no way out—except war—which would remove all the bewildering paradoxes of their tedious and now misguided attempts to construct peace. In place of these paradoxes they prefer the bright, clear problems of war—as they used to be. For they still believe that 'winning' means something, although they never tell us what.[74]

73 Anzalone, Kyle. "General Milley Predicts Grim Future of Deadly Great Power Wars Fought in Cities." *Antiwar.com*, May 22, 2022, https://news.antiwar.com/2022/05/22/general-milley-predicts-grim-future-of-deadly-great-power-wars-fought-in-cities/.
74 Mills, C. Wright, *Op. cit.*, p. 88.

I think that any serious reflection on the post-Cold War era would conclude that it has mainly demonstrated the *futility* of military power. The United States emerged from the Cold War as the greatest military power in the history of the earth, with an unprecedented military superiority over all other countries. To this day, American military spending constitutes 40% of all global military spending. Its budget surpasses that of the next ten countries combined. The British Empire, the dominant power of the 19th century, never had anything like this superiority over other countries. Such was its superiority that American leaders proclaimed it could now remake the world in its image. Such was its superiority that they confidently predicted America would now lead others into an era of peace and prosperity as previous empires had aspired to do—but failed to do, because they lacked the muscle.

What have we seen since?

The United States invaded Afghanistan to overthrow the Taliban. Twenty years later, the Taliban drove the United States out of Afghanistan with a rag-tag army of less than 80,000 soldiers. The United States invaded Iraq with the aim of bringing democracy to the region as part of ending the global terror threat. Just a few years later, it was bogged down by an insurgency that led to larger groups like ISIS that destabilized the whole region. After the Cold War, the power of the United States contrasted starkly with the powerlessness of Russia, whose leaders proclaimed their wish for peace with their former Cold War adversaries. It is not a foregone conclusion that enemies remain enemies after a war. After a bloodier and crueler conflict, World War II, the United States had established peaceful relations of mutual prosperity with Germany and Japan. By contrast, the post-Cold War era looks disturbingly like the decades after World War I, when military victory laid the groundwork for new conflict and new tensions—in which, for all its military power, the United States has proved powerless to prevent.

Had enough?

[V]/ An Outline of What Is to Come

There are other excellent discussions of pacifism and nonviolence. My discussion is most distinguished from them in its structural and historical account of the war system drawing on work of political sociology and history. I have benefited most from the writings of Charles Tilly, Michael Mann, Mary Kaldor, Bruce Porter, Anthony Giddens, Martin Shaw, Niall Ferguson, and Philip Bobbitt on war and the state. Francis McCall Rosenbluth and John Ferejohn have recently provided an excellent overview of war's paradoxical relation to democracy. The historical works on which I have chiefly relied are by Sir Michael Howard, Michael Roberts, William

McNeil, Paul Kennedy, and John Brewer. The impact of world systems theory will be evident in my discussion of early modernity. Fernand Braudel and Immanuel Wallerstein are the reference points here, but my greatest debt is to the work of Giovanni Arrighi and Beverly J. Silver and their merging of the sociological and historical perspectives. My discussion of nationalism is especially indebted to Andreas Wimmer and Benedict Anderson. The works of these scholars are referenced in the text and in the bibliography. For overviews of the literature, I recommend Sinisa Malesevic's *The Sociology of War and Violence* and the article "Towards a Strong Program in the Sociology of War, the Military and Civil Society" by Brad West and Steve Matthewman.[75]

The following chapter introduces my expansive conception of pacifism as a tradition constituted by a dialogue between its major figures. I purposefully discuss some figures that have now been forgotten in the hope of reclaiming them. I then consider why pacifism came to be so maligned over the last century, and then how current developments in just war discussions have made a place for it. Chapter 3 steps back from the historical detail and introduces some general distinctions useful for approaching pacifism as a whole. The principal one is between what I call *personal pacifism* and *political pacifism*. I introduce political pacifism's critique of the war system, and then conclude with some remarks on the problem of "absolutism" and pacifism. A prominent form of pacifism today, contingent pacifism, can be understood as a response to the worry about absolutism. I suggest that the difference between it and the pacifism that I endorse is a methodological one, of whether one appraises wars one at a time or as a system. This is a matter of how one conceives the "practice" of war.

Chapter 4 expands on the critique of the war system via a historical sketch that leads up to the present and concludes with some comments on the political challenges of today. Chapter 5 offers some reflections on questions of traditional and contemporary concern to pacifists like memory, sacrifice, grief, and hope. Special attention is given to issues of gender. They are meant to be suggestive rather than conclusive, contributions to the ongoing dialogue about peace. The Appendix discusses the question of pacifism and self-defense from a variety of perspectives, including how the right to self-defense has been related to war generally.

75 Malesevic, Sinisa. *The Sociology of War and Violence.* Cambridge University Press, 2010; West, Brad West and Mattherman, Steve. "Towards a Strong Program in the Sociology of War, the Military and Civil Society." *Journal of Sociology*, 52(3), 2016, pp. 482–499.

Chapter 2

Pacifism as Tradition

Accounts of pacifism often dwell on its many possible meanings, dissecting all the ways the term has been or might be used. Oftentimes, a ranking is included; that is, there is reference to the "true" pacifist or the "real" pacifist or pacifism "strictly construed," etc. Any such claims are nonsense. They are judgments parading as definitions. There is no "true"/"real" pacifism "strictly construed" any more than there is any "true"/"real" liberalism "strictly construed." Questions could be raised about the whole point of such taxonomies. One might, for example, construct one for a vibrant, dynamic tradition like liberalism; that is, one might list 231 meanings of the term "liberalism," stretching back over centuries and bouncing around over different continents. But the upshot of such a sarcophagal exercise would be to ignore—if not deny—the vibrancy and dynamism of liberalism *as* a tradition. The same holds for pacifism.

Given pacifism's vibrancy and dynamism, any characterization of it will be as much a proposal as a description. I have noted that the term pacifism was originally understood expansively, including much that is now discussed under topics like anti-militarism, critical security studies, and the like. This is how I shall construe it here. There have always been important differences between its variations but they have also overlapped with each other. Plus their proponents have not always placed much importance on their differences with one another, regarding them rather as an occasion for discussion rather than opposition. I shall suggest that the question of how the differences in pacifism came to be seen as so momentous is the question of how pacifism came to be so marginalized in the 20th century. For once "the pacifist" became an object of scorn, it became important for anyone attracted to antiwar-ism to detail his or her relation to it.

[I]/ Pacifism(s)

The term "pacifism" is fairly new, coming from the Latin *pacem facere* for "peace maker." But it can be taken as referring to a tradition of antiwar activism/argument/reflection dating back in the West to first Christianity and

DOI: 10.4324/9781032686189-2

dating further back elsewhere. When first coined, the meaning of "pacifism" was an ecumenical one. It was first proposed in 1901 by French peace campaigner Émile Arnaud and adopted by others at the tenth Universal Peace Congress in Glasgow to replace clumsier terms like "friends of peace." At a time when war was celebrated in many quarters, a "pacifist" was simply someone who rejected pro-war jollifications, who saw war as an intrinsically *bad* thing, and who sought alternatives to it in mechanisms like international arbitration. Self-proclaimed pacifists included mainstream figures and ranged across the political spectrum. They included prominent politicians like presidential candidate William Jennings Bryan, Great Britain's Ramsay MacDonald, French socialist Jean Jaurez, prominent authors like France's Roman Rolland and George Bernard Shaw, and prominent philosophers like William James and Bertrand Russell.

[1]/ "Big Tent" Pacifism

Let us call this expansive pacifism "big tent" pacifism.

Its intellectual origins were secular ones, tracing back to the 18th century Enlightenment and anticipated by humanist figures of the Renaissance like Erasmus. It coalesced in response to the Napoleonic Wars. Peace societies founded in Britain (1816) and the United States (1828) led to an initial series of peace congresses in the 1840s, when the term "peace movement" became common. The first was held in Paris (1849) at the initiative of French author/human rights activist Victor Hugo. Also prominent was American philanthropist and activist Elihu Burritt, who established the League for Universal Brotherhood in 1847. The 1860s saw the emergence of organizations linking peace to causes like socialism and progressive nationalism, such as *The Ligue internationale de la Paix et de la Liberté*. The end of the century witnessed a series of international peace congresses calling for the establishment of a "Permanent International Peace Bureau" to coordinate such efforts, all of which coincided with the first official efforts at arms control and international arbitration.

A representative voice of such "big tent" pacifism was Austrian-Bohemian Bertha von Suttner (1843–1914). Her book *Das Maschinenzeitalter* [*The Machine Age*] was among the first to foretell the results of inflamed nationalism and armaments. She followed it with her international bestseller, *Die Waffen nieder!* [*Lay Down Your Arms!*]. Written in an autobiographical style, it sought to convey on an experiential level the full horrors of war. Published late in 1889, its depictions were so real and its indictment of militarism so telling that its impact on the reading public was akin to *Uncle Tom's Cabin's* on slavery. She went on to be a leader in the peace movement, becoming in 1905 the first woman to win the Nobel Peace Prize "for her audacity to oppose the horrors of war." The role of women was a prominent theme. She wrote in *International Peace Through the Voice of Women* (1912) that "Through the voice of the women will come international peace. Not until that voice is raised will the federation of the world in brotherly love be accomplished."[1] Her Nobel lecture struck an optimistic note: "Quite apart from the peace movement, which is a symptom rather than a cause of actual change, there is taking place in the world a process of internationalization and unification." The peace movement "had emerged from the fog of pious theories into the light of attainable and realistically envisaged goals." Still, "pacifism faces no easy struggle."[2]

Writing on the eve of World War I, she stressed that the cause of peace would determine "whether our Europe will become a showpiece of ruins and failure, or whether we can avoid this danger" to create "a civilization of unimagined glory." She died in June 1914, just as her worst fears for Europe were being realized.

Just as World War I was the formative event of the 20th century, it was the decisive moment for the politics of peace. Theoretically, pacifism became more expansive in its analysis of the "war system." The growth of the social sciences in the late 19th century prompted more systematic studies of war. Along with the writings of Jan Gotlib Bloch, Norman Angell, and Randolph Bourne, already noted, Austrian Alfred Freid had called for a "scientific pacifism" to explore the organizational bases of war. Culturally, the war inspired a literature of disillusionment, especially among those who had fought in it that laid the basis for what I have elsewhere termed the "soldier-pacifist tradition" of those whose first-hand experiences of war led them to reject all war.[3] Practically, the war saw the founding of such groups as the War Resisters League in the Netherlands in 1921 by such figures as Helene Stöcker, whose pacifism would lead to her heroic opposition to the Nazis, and Bartholomeus de Ligt, whose writings on pacifism and anarchism would become influential in the 1930s.

1 Von Suttner, Bertha. "International Peace Through the Voice of Women." *Archives of Women's Communication. Iowa State University,* July 2, 1912. https://awpc.cattcenter. iastate.edu/2018/10/18/international-peace-through-the-voice-of-women-july-2-1912/.

2 Von Suttner, Bertha. "The Evolution of the Peace Movement." *Nobel Lecture,* April 18, 1906. www.nobelprize.org/prizes/peace/1905/suttner/lecture/.

3 Ryan, Cheyney. "The Lament of the Demobilized." *To End a War: Essays on Justice, Peace, and Repair.* Edited by Graham Parsons and Mark Wilson, Cambridge University Press, 2022.

[2]/ Anti-Imperialist Pacifism

For some, World War I solidified the connection between the war system and the capitalist system. This was the theme of those left-wing socialist who had refused to go along with the general acquiescence of the socialist movement to the war frenzy. They certainly did not renounce violence; on the contrary, they saw ending war as achievable through class violence. But their analysis of war and class and their insistence that workers potentially constituted a transnational community against war became significant features in the critique of the war system.

Karl Liebknecht's *Militarism and Anti-Militarism* (1907) earned him his first prison sentence from German authorities; his second was for his role in the 1916 antiwar demonstrations. "Militarism!" he wrote, "Few slogans have been so frequently used in our time, and few denote a phenomenon so complicated, multiform, many-sided, and at the same time so interesting and significant in its origin and nature, its methods and *effects*." He continued,

> The history of militarism is at the same time the history of the political, social, economic and, in general, the cultural relations of tension between states and nations, as well as the history of the class struggles within individual and national units.

Hence, a challenge to the working class "will be the struggle against militarism in all its aspects."[4] In her "The Meaning of Pacifism" (1911), Rosa Luxemburg voiced full support for the general aims of the peace movement while rejecting that "world peace and disarmament can be realized within the framework of the present social order." The need was for more radical solutions.

> Militarism in both its forms—as war and as armed peace—is a legitimate child, a logical result of capitalism, which can only be overcome with the destruction of capitalism, and hence whoever honestly desires world peace and liberation from the tremendous burden of armaments must also desire socialism.[5]

---

4 Liebknecht, Karl. Militarism and Anti-Militarism: *With Special Regard to the International Young Socialist Movement*. Translated by Grahame Lock, Rivers Press Limited, 1973.

5 Luxemburg, Rosa. "The Meaning of Pacifism." *Socialist Appeal*, Vol. II No. 14, April 2, 1938. www.marxists.org/history/etol/newspape/themilitant/socialist-appeal-1938/v02n14/luxemburg.htm.

Both Liebknecht and Luxemburg participated in the left-wing uprising at the end of World War I that was part of the general domestic turmoil that ended the war. Government-sponsored paramilitary groups murdered them both during the rebellion.

Perhaps the most eloquent voice of socialist anti-militarism was America's Eugene Debs. Debs came to prominence in the 1890s when he led a nationwide boycott in support of railroad workers striking against the Pullman Car Company. It led to a six-month prison term during which reflection and reading turned him into a committed socialist. He participated in 1905s founding convention of the International Workers of the World, and his passion and eloquence made him the most prominent socialist voice of his era. More than a million people read his newspaper *Appeal to Reason*, the equivalent of more than three million people today. Like Liebknecht and Luxemburg, he did not acquiesce to the pro-war sentiments of other socialists when the conflict began in Europe. This would eventually lead to his imprisonment when President Woodrow Wilson instituted draconian laws against free speech as part of his general mobilization for the war effort.

What got him in trouble was a speech in Canton, Ohio, in support of men and women in prison for protesting the war. That, and his statement to the court upon being sentenced to ten years in prison for violating the Sedition Act, remain two of the most eloquent antiwar testimonies in the history of American activism. "Wars throughout history have been waged for conquest and plunder," he stated in his Canton speech, "And that is war, in a nutshell. The master class has always declared the wars; the subject class has always fought the battles."

"They are continually talking about your patriotic duty. It is not their but your patriotic duty that they are concerned about. There is a decided difference. Their patriotic duty never takes them to the firing line or chucks them into the trenches." "They have always taught and trained you to believe it to be your patriotic duty to go to war and to have yourselves slaughtered at their command. But in all the history of the world you, the people, have never had a voice in declaring war, and strange as it certainly appears, no war by any nation in any age has ever been declared by the people."[6]

6 Debs, Eugene. Canton Speech. *Chicago: Socialist Party of the United States*, 1918. https://college.cengage.com/history/ayers_primary_sources/eugene_cantonspeech_1918.htm.

He ran for president in 1920 while imprisoned in a federal penitentiary, receiving almost a million votes. After the war, President Wilson refused to pardon him despite the repeated recommendation for clemency from his Attorney General. By contrast, the newly elected President Warren Harding commuted his sentence in December 1921. He was nominated in 1924 for the Nobel Peace Prize by Finnish socialist Karl H Wiik on the grounds that "Debs started to work actively for peace during World War I, mainly because he considered the war to be in the interest of capitalism."[7]

[3]/ Anti-Colonialist Pacifism

The West develops wonderful new skills,
In this as in so many other fields
Its submarines are crocodiles
Its bombers rain destruction from the skies
Its gasses so obscure the sky
They blind the sun's world-seeing eye.
Dispatch this old fool to the West
To learn the art of killing fast—and best.
—Muhammad Iqbal, *A Message
from the East*[8]

Over 200,000 Chinese laborers were conscripted by the Allies to fight on the European front. France enlisted 100,000 Vietnamese peasants to that end. At the start of World War I, the Indian Army was equivalent in size to the British Army. The one million Indians who fought for the British served in Europe, Egypt, Gallipoli, and German East Africa. By far the largest contingent—nearly 700,000—served in Mesopotamia initially to guard British oil installations. On the whole, over 30,000 of them died from wounds or disease. Africa was a major scene of conflict as well, principally over the German colonies and involving forces from all the major combatants. But the principal casualties were Africans themselves, many of whom were forced to fight their own families because borders drawn by Europeans were so arbitrary. By far the greatest losses were among Africans serving as "carriers" or other forms of support for the troops. Entire areas of Africa were depopulated from such enlistments. Of the

7 The Nomination Database for the Nobel Prize in Peace, 1901–1955. https://web.archive. org/web/20071012203613/http://nobelprize.org/nomination/peace/nomination.php?actio n=show&showid=1347.

8 Mishra, Pankaj. *The Ruins of Empire: The Revolt against the West and the Remaking of Asia*. Reprint ed. Penguin Books Ltd, 2013, p. 210.

one million Africans so enlisted by the British in East Africa, around 10% perished. The resulting demographic and economic devastation had catastrophic consequences for the continent. Famine resulted in over 300,000 civilian deaths in German East Africa, 200,000 deaths in British East Africa, and over 300,000 deaths in South Africa.

Japan's victory over Russia in the Battle of Tsushima (1905) had punctured the myth of European invincibility. A decade later, World War I would undermine Europe's moral prestige in non-Western eyes. Raymond Aron wrote, "Europeans would like to escape from their history, a 'great' history written in letters of blood. But others, by the hundreds of millions, are taking it up for the first time, or coming back to it."[9] One result was the most expansive challenge to Western modernity as an enterprise of violence. European socialists focused on the economic dimensions of the war system primarily. The anti-colonialist perspective, as I am calling it, extended the critique to every dimension of social life, especially the spiritual one. In the words of Rabindranath Tagore, "[T]he torch of European civilization was not meant for showing light, but to set fire."[10]

Its most well-known voice remains Mohandas Gandhi (1869–1948), who did the most to enlarge the pacifist tradition in the 20th century. The depth of his influence is marked by considering two of his followers.

"The Frontier Gandhi": this was the name given to one of the more significant but lesser-known figures in nonviolence, Pakistan's Abdul Ghaffar Khan (1890–1988). Though called "the frontier Gandhi," his ideas about nonviolence and social change developed independently and were based on his understanding of Islam, whose essence he identified with peace. His ability to mobilize his Pashtun people was truly extraordinary. He organized what was a nonviolent army of over 80,000 men and women, the Khudai Khidmatgars (The Servants of God), whose members pledged "to refrain from violence and from taking revenge. I promise to forgive those who oppress me or treat me with cruelty." They engaged in nonviolent resistance against the British and promoted other forms of social change including religious tolerance and land distribution. Not surprisingly, he was despised by the British, who regarded his commitment to nonviolence as a sham. His followers were subjected to mass killings, torture, and

---

9 Aron, Raymond. *D'Une Sainte Familla a l'autre*, p. 13. Cited in Judt, Tony. *The Burden of Responsibility: Blum, Camus, Aron, and the French Twentieth Century*. University of Chicago Press, 2008, p. 158.

10 Chung, Tan, et al., editors. *Tagore and China*. 1st ed. SAGE Publications Pvt. Ltd, 2011, p. 79; Chand, Mool. *Nationalism and Internationalism of Gandhi, Nehru, and Tagore*. M.N. Publishers and Distributors, 1989; Xu, Guoqi. *Asia and the Great War: A Shared History*. 1st ed. Oxford University Press, 2017.

destruction of their homes and fields throughout the 1930s and 1940s. Khan himself spent over 15 years in prison.[11]

"The Japanese Gandhi": this is how Japan's Toyohiko Kagawa (1888–1960) has been described. He was well known in the United States at one time, but is now largely forgotten. His fame in America partly reflected his commitment to Christianity, which led some in the Christian community to champion him as a Christian Gandhi as well. Like Gandhi, he was deeply involved in the issue of poverty which led him to lifelong engagement with the urban poor. In 1928, he organized the National Antiwar League to counter the rise of Japanese militarism which he worked to oppose for the rest of his life. He was arrested in 1940 after apologizing to the Republic of China for Japan's occupation of it, after which he traveled to the United States in a futile attempt to prevent war between his country and the United States. As an organizer of the Japanese Federation of Labor, he remained committed to the notion that movements of economic cooperation should be joined to the peace movement as a "powerful working synthesis" as an alternative to capitalism and communism as well as war.[12]

Finally, there is the magisterial figure of India's Rabindranath Tagore (1861–1941). Tagore's achievements gave new meaning to the word polymath. In the course of his life, he made major contributions to literature, music, and art, as well as being a major figure in the reform of education. He is often said to have reshaped Bengali literature. He increasingly turned to peace activism after winning the Nobel Prize for literature in 1913, which assured his prominence in Western eyes. In a 1914 Vienna speech (at the invitation of Bertha von Suttner), Tagore spoke of how war was the product of aggressive Western materialism and its divorcing of science from Spirituality. "A new readjustment of things is necessary, a new age, when the idea of nationalism will be discarded, when colonies, the storm centres of the world, will be discarded."[13]

Tagore is representative of the struggles over the meaning of nationalism that characterized anti-colonialist movements, in particular, over the

---

11 Sahibzada, Imitiaz Ahmad. *The Frontier Gandhi: My Life and Struggle: The Autobiography of Abdul Ghaffar* Khan. Roli Books, 2021; Johansen, Robert C. "Radical Islam and Nonviolence: A Case Study of Religious Empowerment and Constraint among Pashtuns." *Journal of Peace Research*, 34(1), 1997, pp. 53–71; Sharify-Funk, Meena. "Toward a Global Understanding of Pacifism: Hindu, Islamic, and Buddhist Contributions." *Pacifism's Appeal: Ethos, History, Politics*. Edited by Jorg Kuystermans and Tom Sauer and Dominiek Lootens and Barbara Segaert. 1st ed. Palgrave Macmillan, 2019.

12 Schildgen, Robert. *Toyohiko Kagawa: An Apostle of Love and Social Justice*. 1st ed. Centenary Books, 1988.

13 Speech by Tagor, cited in Kundu, Kalyan. "Rabindranath Tagore and World Peace." *Asiatic*, 4(1), 2010, p. 81.

extent to which nationalism *per se* was implicated in war. He denounced European imperialism and supported anti-colonialist movements. He resigned the knighthood that had been conferred on him in protest of British soldiers massacring unarmed Indian civilians. On the other hand, he denounced nationalism as among humanity's greatest evils. "A nation," he wrote, "is that aspect which a whole population assumes when organized for a mechanical purpose." A "mechanical" purpose, for him, was a spiritually vacuous one, selfishness transmuted into magnified form.[14]

[4]/ Feminist Pacifism

The figures so far might be rightly called political ones. But pacifism's voice has taken other forms, as its engagement with feminism illustrates. Here, there are what would mainly be considered literary figures but whose opposition to war was often central to their efforts. I think Mary Wollstonecraft would be included here for her reflections on war and masculinity, and Virginia Wolf in writings such as *Three Guineas* (1938).[15] For my generation, Murial Rukeysar and Adrienne Rich were essential voices of this sort, and more recently Carolyn Forche.

World War I was also the defining moment for feminist pacifism, marked by the heroic resistance of women's organizations to that conflict. Feminism and peace were already linked in the late 19th century. In the United States, what is now celebrated as an a-political "Mother's Day" originated from this. Julia Ward Howe had issued a "Mother's Day Proclamation" calling on mothers of all nations to promote the "amicable settlement of international questions." She was later joined by Anna Jarvis, moved by her nursing experiences in the Civil War, to urge a "Mother's Day For Peace" where mothers would urge that their husbands and sons should be no longer sent to war. Late 19th-century socialists had regarded workers as the natural community for peace, but that vision was shattered by mainstream socialism's support for World War I. Feminists now argued that women constituted the international audience for pacifism and they soon became the backbone of organizational forces against the war.

January 1915 saw 3,000 women participating in the foundation of the Woman's Peace Party in Washington, DC. Its platform was a robust one. It called for arms limitation, legislation, and economic policies to prevent war, the replacement of national militaries by an international police force, international institutions to resolve conflicts between nations, the

14 Tagore, Rabindranath. *Nationalism*. MacMillan and Co. Ltd, 1918, p. 9.
15 Andrew, Barbara. "The Psychology of Tyranny: Wollstonecraft and Woolf on the Gendered Dimension of War." *Hypatia*, 9(2), 1994, pp. 85–101.

mediation of the conflict that had just broken out, and the central involvement of women in all these activities.

At the center of all these activities was Jane Addams (1860–1935), America's leading peace activist in the first half of the 20th century and one of the world's leading peace voices during that time. Her book *Newer Ideals of Peace* had argued for an expansive peace movement that included other issues of social justice as well. After her election as chairwoman of the Woman's Peace Party, she was invited to preside over an International Congress of Women in The Hague in April 2015. Its organizers included Aletta Jacobs (the Netherlands), Gabrielle Duchene (France), Marianne Cripps (Great Britain), and two Germans who have been called the "daughters" of Bertha von Suttner—Anita Augbpurg and Lida Gustava Heymann. Heymann's call echoed Suttner's words:

> Women of Europe, where is your voice? Come together in the north or south of Europe, protest strongly against this decimating war while preparing for peace, return to your own lands and repeat the call, fulfill your duty as women and mothers, as caretakers of true culture and humanity.[16]

Over 1,000 women from across Europe and all the war's contending powers gathered to call for an end to the conflict and permanent institutions to end war. Following the conference, Addams and other leaders engaged in an extraordinary journey throughout Europe to meet with leaders, citizen groups, and wounded soldiers from all sides to press the cause of reconciliation. It was the most significant effort of such a sort during the war. Afterward, an International Committee of Women for a Permanent Peace was established in 1919 to continue the work of the Hague Conference, which subsequently became the Women's International League for Peace and Freedom (WILPF)—still an important antiwar group. Jane Addams was awarded the Nobel Peace Prize in 1932, only the second woman to receive it.

Another prominent voice for feminist pacifism was Britain's Vera Brittain (1897–1970), whose *Testament of Youth* (1933) recounting her wartime experiences became an enduring classic of pacifism. Trained as a nurse, she volunteered for overseas service in 1916 requesting posting to the Western Front where she remained until 1918. Her time there

16 Braker, Regina. "Bertha von Suttner's Spiritual Daughters: The Feminist Pacifism of Anita Augspurg, Lida Gustava Heymann, and Helene Stöcker at the International Congress of Women at the Hague, 1915." *Women's Studies International Forum*, 18(2), 1995, pp. 103–111; 108.

included nursing German prisoners of war, which she recounted as having a profound impact on her aversion to war. Her fiancé, Roland Leighton, her close friends Victor Richardson and Geoffrey Thurlow, and finally her brother Edward were all killed in the war. Her literary success led to her involvement with groups like the Peace Pledge Union, the largest pacifist organization in Britain, while her involvement with the Anglican Pacifist Fellowship marked her increasing inspiration by Christianity. She became a committed pacifist after reading Bertrand Russell's *Which Way to Peace?* (1936), which concluded: "modern war is practically certain to have worse consequences than even the most unjust peace." She worked to establish "a higher order of human relationships in which war is no more, the exploitation of man by man unknown, and the subjection of women a cruelty of the past."[17] She continued to speak out in the 1950s and 1960s about nuclear disarmament, as well as issues like apartheid and colonialism.[18]

[5]/ Nonresistance

By "nonresistance," I principally mean the personal refusal to engage in killing. I prefer this to other definitions, though others are equally valid. One is the refusal to engage in all violence; this raises larger questions of how to understand "violence." Another is the refusal to resist all authority. This places nonresistance in *opposition* to nonviolence, as a way of resisting authority, which strikes me as unhelpful in understanding how the tradition has unfolded.

This perspective is religious in origins, with all the major religions having variations, albeit marginal ones, endorsing nonresistance. After the early Christians, it first became a distinctive orientation in the West during the "Radical Reformation." This was the 16th-century response to the corruption in the Catholic Church as well as the perceived corruption of the mainstream Protestantism of Martin Luther and others. It gained special importance in Germany, Switzerland, and Austria and would later be influential in the United States. Its principal form was Anabaptism, so named for its opposition to infant baptism from the belief that baptism was only valid when the baptized could confess their faith. The Amish, Hutterites, and Mennonites are its principal descendants, who generally base their views on a literal interpretation of the Sermon on the Mount and its injunction

17 Brittain, Vera. *Thrice a Stranger*. Victor Gollancz Ltd, 1939, p. 21.
18 For an overview, see *The Oxford Handbook of Gender, War, and the Western World since 1600*. Edited by Karen Hagemann, Stefan Dudink and Sonya O. Rose, Oxford University Press, 2020.

to turn the other cheek. Thomas Munzer stated in 1524 that "True Christian believers are sheep among wolves, sheep for the slaughter . . . Neither do they use worldly sword or war, since all killing has ceased with them."[19] Since they have generally held themselves apart from civil government, their political impact has been negligible. Anabaptism's importance today is its place in the thinking of the two major Anglophone theologians of pacifism, John Howard Yoder and Stanley Hauerwas.

Nonresistance became politically impactful in the next century in England during its civil war. Its founder, George Fox (1624–1691), was in and out of prison for preaching a Christianity that, while biblically oriented, relied more on personal experience than the written word. The first explicit statement of Quaker nonviolence was by James Naylor in his book, *The Lamb's War* (1657). He argued that violence was rooted in anger that arose from the fundamental state of sin; hence, we all carry evil within us. If we can purge ourselves of pride, we will render ourselves incapable of committing violence. A few years later, Quakers officially adopted nonresistance as a fundamental principle for the Society of Friends. The story of how Quakerism gained political influence is partly a matter of how it survived the English Civil War as other radical sects did not, undoubtedly because they were not as radical—they did not condemn private property, for example. But Quakers also took seriously the calling to make the world more "godly" by "overcoming evil with good."

Nonresistance's ethics is illustrated in the life of John Woolman (1720–1772), a key figure in the transmission of Quaker ideas to America. His writings, such as his *Journal* and essays like "A Plea for the Poor" were widely read and admired well into the 18th century. Fox had made his aim to revive the spirit of Christianity by living in harmony with the Sermon on the Mount. But he did not construe this in a legalistic way. The essence of the religious life was "sensitiveness" to the divine light in others that both presumed and nurtured the divine light within oneself. Woolman's *Journal* impressed readers as testimony to this practical love, which was meant to pervade one's whole being and lead to the renunciation of violence in any form. "It requires great self denial and resignation of ourselves to God, to attain that state wherein we can freely cease from fighting when wrongfully invaded, if, by are fighting, there were a probability of overcoming the invaders."[20] The Quakers were notable for the expansiveness of their ethic that counseled more respectful relations with Native Americans. In Woolman's hands, its ethic involved a deep suspicion of the possessiveness

19 Dyck, Cornelius. *An Introduction to Mennonite History*. Herald Press, 1967, p. 45.
20 Woolman, John. *The Journal of John Woolman and A Plea for the Poor*. Wipf and Stock, 1998, p. 77.

associated with private property in closing us off to others. Woolman stressed the link between the evils of war and the evils of slavery which he regarded as both arising from the sin of greed. "We may we look upon our treasures, the furniture of our houses, and our garments, and try whether the seeds of war have nourishment in these our possessions," he wrote in "A Plea for the Poor." "Wealth is attended with power, and hence oppression, carried on with worldly policy and order, clothes itself with the name of justice and becomes like a seed of discord in the soul." He relates "the calamities occasioned by contending Nations," for which "selfishness has been the original cause of them all," to the fact that those who are "afflicted" by such selfishness "have no real quietness in this life nor in futurity."[21] Inner harmony, as consistency with principle, leads to the work of political peace, but its ground and reference points remain a kind of inner peace.

Nonresistance came into its own as a political force with the 19th-century American abolitionist movement against slavery. Abolitionism was the major American radical movement of the 19th century, rivaled only in American history by the Civil Rights Movement of the 20th century. It too was a "big tent" movement driven by different ideologies and directed toward different strategies. But its most powerful voice was William Lloyd Garrison (1805–1879), who would also become America's leading pacifist of the 19th century.[22]

For Garrison, the connections he saw between slavery and the armed state made the journey from abolitionism to pacifism a natural one. Like other abolitionists, he regarded slavery as essentially a form of war. But while he was a committed Christian and invoked Christianity to articulate his views, his pacifism was not theologically parochial; on the contrary, it was grounded in specific experiences that led him to become an activist against war as well as slavery.

Garrison was notable for the inclusiveness of his politics, provoking conflict with other abolitionists in the prominent role he accorded women. Having just addressed a Boston meeting of a women's anti-slavery society, he was confronted by an angry mob, inflamed by proslavery denunciations of him. He truly believed he would be killed and was only spared by a well-meaning magistrate who placed him in jail for his own safety. His abolitionism was not altered, but it proved to be a defining moment for his pacifism, which he identified with his equanimity in facing death. It confirmed for him, he wrote later, that the most transforming influence on

---

21 *Ibid.*, 239–241.
22 The best introduction to Garrison is Mayer, Henry. *All on Fire: William Lloyd Garrison and the Abolition of Slavery.* 1st ed. St Martin's Press, 1998.

the world was the endurance by people of principle of barbarous treatment "without resorting either to their own physical energies, or to the force of human law, for restitution or punishment."[23] To a fellow abolitionist who urged a violent response, he wrote,

> Do you wish to become like one of those pilots and bloodthirsty man who are seeking my life? . . . I will perish sooner than raise my hand against any man, even in self-defense, and let none of my friend's resort to violence for my protection.[24]

Garrison's New England Non-Resistance Society was founded in 1838, along with its journal "Nonresistant." At the time, he predicted that its founding would rival July 4 as the herald of a new epoch of humanity. The organization was notable for its uncompromising opposition to all war. He wrote in its founding "Declaration of Sentiments,"

> We register our testimony, not only against all war, whether offensive or defensive, but all preparations for war; against every naval ship, every arsenal, every fortification; against the militia system and a standing army; against all military chieftains and soldiers; against all monuments commemorative of victory over a foreign foe, all trophies won in battle, all celebrations in honor of military or naval exploit; against all appropriations for the defense of the nation by force and arms, on the part of any legislative body; against every edict of government requiring of its subjects military service.

To which he added:

> We can allow no appeal to patriotism, to revenge and a national insult or injury. The Prince of peace, under whose stainless banner we rally, came not to destroy, but to save, even the worst of enemies. He left us as an example, that we should follow the steps.[25]

---

23 *Ibid.*, p. 222.
24 Garrison, William Lloyd. *William Lloyd Garrison on Non-Resistance, together with a Personal Sketch by His Daughter Fannie Garrison Villard and a Tribute by Leo Tolstoy.* The Nation's Press Co., 1924, p. x.
25 Garrison, William Lloyd. *William Lloyd Garrison and the Fight Against Slavery: Selections from the Liberator.* Edited by William Cain. 1st ed. Bedford/St. Martin's Press, 1994, p. 102.

Finally, his method was that of moral witness, which he saw as necessary for the kind of general moral transformation that the renunciation of war would involve. He wrote,

> Moral influence, when in vigorous exercise, is irresistible. It has an immortal essence. It can no more be trod out of existence by the Iron foot of time, or by the ponderous march of iniquity, that matter can be annihilated. It may disappear for a time; but it lives in some shape or other, in some place or other, and will rise would renovated strength.[26]

The pacifism of important figures is often marginalized if not ignored entirely in retrospect. Jane Addams is now remembered as an amiable social worker; William Lloyd Garrison, if remembered at all, is done so only for his abolitionism. But one who recognized Garrison's importance and acknowledged him as a precursor of his own views was Russian novelist Leo Tolstoy (1828–1910), who cites him at length at the start of his pacifist classic *The Kingdom of God is Within You* (1894).

Tolstoy is the towering figure of nonresistance in the late 19th and early 20th centuries.[27] He was appalled by his experiences in army service in the Crimean War. That and his later experiences in Europe, including witnessing a public execution in Paris, transformed him from a scion of privileged society to a nonviolent and spiritual anarchist. His was a Christian-based doctrine, but he regarded Christianity and specifically Jesus' teaching as capturing universal moral commitments. In Quaker fashion, he believed that God had placed a divine light in every man to which the Sermon on the Mount gave voice but which every religion was capable of recognizing. Christ's teachings "have now become identified with human conscience," and the voice of conscience "is also the voice of reason."

His practical approach was one of moral persuasion, specifically calling people to acts of individual resistance. He wrote that ending war would be a matter of conscience more than politics: "the easiest and surest way to universal disarmament is by individuals refusing to take part in military service."[28] Hence, his faith in the power of moral witness.

> Just as a single shock may be sufficient, when a liquid is saturated with some salt, to precipitate it at once in crystals, a slight effort may be

---

26 Ibid., p. 67.

27 Christoyannopoulos. Alexandre. *Tolstoy's Political Thought: Christian Anarcho-Pacifist Iconoclasm Then and Now.* 1st ed. Routledge, 2021.

28 Tolstoy, Leo. "Letter on the Peace Conference." *Tolstoy's Writings on Civil Disobedience and Nonviolence.* New American Library, 1968, pp. 113–119; 113.

perhaps all that is needed now that the truth already revealed to men may gain a mastery over hundreds, thousands, millions of men, that a public opinion consistent with conscience may be established, and through this change of public opinion the whole order of life may be transformed. And it depends upon us to make this effort.[29]

Tolstoy's type of pacifism is often regarded, and dismissed, as hopelessly "non-pragmatic." Tolstoy himself renounced any desire to form a movement. He remarked,

To speak of "Tolstoyism," to seek guidance, to inquire about my solution of questions, is a great and gross error. There has not been, nor is there any "teaching" of mine. There exists only the one eternal universal teaching of the Truth, which for me, for us, is especially clearly expressed in the Gospels.

But his "teachings" nevertheless had profound impact in his time. Gandhi named his cooperative colony after Tolstoy. Tolstoy-inspired communes existed throughout Russia up to 1921 when the Bolsheviks abolished them. They endured in England and the Netherlands.

The doctrines of Tolstoy—"Tolstoyism"—began circulating in the United States in the second half of the 1880s. In 1889, a Tolstoy club was founded in Boston, committed to causes like agrarianism, vegetarianism, and abstinence from alcohol. His most prominent American follower was Ernest Edward Crosby, a politician-turned judge who resigned his position after reading Tolstoy to journey to Russia to meet him and advocate for his views throughout the United States. Tolstoy's articles appeared frequently in American journals up to 1917 when they and the larger movement they inspired became victims of pro-war hysteria.

A sign of Tolstoy's influence was the devotion of the renowned civil libertarian attorney Clarence Darrow to his cause. Darrow read Tolstoy in the late 19th century and for the next 20 years established himself as one of America's premier authorities on him by giving lectures across the country. Darrow was especially impressed with the condemnation of retributive justice found in Tolstoy's novel *Resurrection*. In 1903, he published a book, *Resist Not Evil*, which included not just a critique of the prison system but

---

29 Tolstoy, Leo. *The Kingdom of God Is Within You*. Translated by Constance Garnett, independently published, 2021, p. 201.

also a critique of states, militaries, and war generally as all part of the same system. It began,

> It is not claimed that the following pages contain any new ideas. They were inspired by the writings of Tolstoy, who was the first, and in fact the only, author of my acquaintance who ever seemed to me to place the doctrine of non-resistance upon a substantial basis.

His writings were the antidote to the libraries of books that otherwise extolled "the glories of war and the beneficence of patriotism."[30] He acknowledged that in many quarters

> The doctrine of non-resistance, if ever referred to, is treated with derision and scorn. At its best the doctrine can only be held by dreamers and theorists, and can have no place in daily life. Every government on earth furnishes proof that there is nothing practical or vital in its teachings.

But in truth, "Hatred, bitterness, violence and force can bring only bad results—they leave an evil stain on everyone they touch. No human soul can be rightly reached except through charity, humanity and love."[31]

[6]/ Nonviolence

Twentieth-century nonviolence had many precursors. Its principal architects pre-World War II were Gandhi and his followers already noted. I shall focus on post-World War II and the American Civil Rights Movement whose significance includes the social struggles it inspired which continue to this day. Key parts of this legacy include the South African resistance to apartheid, the 1989 popular uprisings against communism, and the nonviolent actions of Palestinians, among others. If one thinks of nonviolence as a distinctive form of 20th–21st century politics, then it is notable for being the only one whose major figures have been mainly non-European (e.g., Gandhi, Abdul Ghaffar Khan, and Thich Nhat Hanh) persons of color (e.g., Dr. King and Cesar Chavez), or women (e.g., Dorothy Day, Ella Baker, and Barbara Deming).

The story of nonviolence's migration from Gandhi in India to the post-World War II African American struggles is one of the more interesting

---

30 Darrow, Clarence. *Resist Not Evil*. Ludwig Von Mises Institute, 2007, p. xi.
31 *Ibid.*, p. 83.

stories of modern politics, if insufficiently known.[32] It culminated Dr. Martin Luther King, Jr. but involved some of the giants of twentieth-century African American politics such as theologian Howard Thurman, college president Benjamin Mays, activist Bayard Rustin, and activist/minister James Lawson. From the start, the African American community saw Gandhi's movement as speaking to their own conditions, which many construed as a form of colonial oppression. And they were attracted to how Gandhi's nonviolent politics was attentive to local traditions, especially religious ones, and to how it offered a wide choice of tactics for confronting problems at hand. It contrasted with other leftist programs that were perceived as imposing aims and solutions from the outside. African American newspapers regularly reported on the Indian independence movement, and a significant number of Black leaders journeyed to India to witness it firsthand (a major task in the days of steamships!).

In 1935, Thurman, the 20th century's most important African American theologian, made a "pilgrimage of friendship" to India to meet Gandhi, the upshot of which he said changed his life forever. Gandhi urged Thurman to develop a new brand of American Christianity to stress social justice themes. He remarked, "[I]t may be through the Negroes that the unadulterated message of nonviolence will be delivered to the world." Others that followed included Benjamin Mays, later president of Morehouse College, whose 1936 speech to students introduced a young Martin Luther King, Jr. to nonviolence, and James Lawson, recently released from prison for draft resistance. The brand of American Christianity they developed would be deeply indebted to the prophetic books of the Hebrew Bible and the prophetic elements of the Christian Gospels. This marks the greatest difference between American nonviolence and that of Gandhi, or Eastern figures like Thich Nhat Hanh. Intellectually, it is evidenced in the powerful influence Martin Buber had on Dr. King's thinking, most clearly in his "Letter from a Birmingham Jail." Politically, it was evidenced in King's strong personal bond with Rabbi Abraham Joshua Heschel, whose masterpiece, *The Prophets*, Dr. King carried with him the day he was assassinated. Though Dr. King preferred to speak of his perspective as nonviolence, he insisted that it constituted in his words "true pacifism." His life embodied the link between the struggle for racial equality and the struggle

---

32 Kapoor, Sudarshan. 1992. *Raising Up a Prophet: The African-American Encounter with Gandhi.* Beacon Press, 2012; Slate, Nico. *Colored Cosmopolitanism: The Shared Struggle for Freedom in the United States and India.* Harvard University Press, 2017; Thurman, Howard. *With Head and Heart: The Autobiography of Howard Thurman.* New Mariner Books, 1981; Ryan, Cheyney. "Bearers of Hope: On the Paradox of Non-Violent Action." *The Ethics of Soft War.* Edited by Michael Gross and Tami Meisels, Cambridge University Press, 2017.

for peace, one that marked 19th century's abolitionist movement. At the time Dr. King was assassinated, he was America's most prominent opponent of the Vietnam War, insisting even against others in his organization that justice and peace could not be separated.[33]

In fact, the historical movement was from antiwar-ism to civil rights then back to antiwar-ism. This is another little-known tale recently recounted in David Akst's *War by Other Means: How the Pacifists of World War II Changed America for Good*. No one was more marginalized in their time than the pacifists who refused to endorse America's efforts in World War II. No one would have imagined that this group, especially its members who went to prison for resisting military service, could have any long-term political impact on anyone. "No rational person observing that movement during the 1940's would have predicted any success at all, and yet during the next two or three decades [these pacifists] transformed whole areas of American life."[34] Akst recounts how A.J. Muste, David Dellinger, Dwight McDonald, and Dorothy Day, and lesser-known pacifist figures like James Farmer, Bernice Fisher, and George Houser had an outsized impact both on specific movements, like the anti-nuclear movement, and on the political/culture orientation of the New Left generally. In his words, "The resisters' loathing of hierarchy, their suspicion of the modern state, and their keen devotion to personal autonomy and moral hygiene would shape the character of the Left into the twenty-first century."[35]

An exemplary figure here was Bayard Rustin (1912–1987). From the influence of his grandparents, Rustin became involved with the Quakers as a young adult. This led to participation in peace activities organized by the American Friends Service Committee (AFSC) in the late 1930s. He would write, "I never would have come to certain social concerns had I missed the experience with the AFSC." In 1942, he assisted in the formation of the Congress of Racial Equality (CORE), supported in its infancy by Muste's pacifist Fellowship of Reconciliation (FOR), which looked to the writings of Gandhi and the nonviolent movement in India as a model for social change in the United States. He traveled to California to help protect the property of the more than 100,000 Japanese Americans imprisoned in internment camps. But during this time he was also convicted of refusing the draft, so from 1944 to 1946 he was in federal prison where he organized

33 In his first book, *Stride Toward Freedom*, Dr. King spoke of nonviolent resistance as "true pacifism." King, Martin Luther. *Stride Toward Freedom*. Ballantine, 1958, p. 80.

34 Akst, Daniel. *War by Other Means: How the Pacifists of World War II Changed America for Good*. Melville House, 2022, p. xii.

35 *Ibid.*, p. xx.

against racially segregated housing and dining facilities as well as finding time to organize FOR's Free India Committee.[36]

In 1948, Rustin traveled to India to learn about nonviolent politics from the leaders of the Gandhian movement at a conference that had been organized prior to Gandhi's assassination earlier that year. He also met with leaders of the African independence movement. This immediately followed his involvement in CORE's "Journey of Reconciliation" where a multiracial group of activists embarked on a two-week bus journey challenging segregation in interstate travel. When Rustin refused to move to the segregated part of the bus, he was severely beaten and arrested. This was the direct precursor of 1961's "Freedom Rides," also organized by CORE.

Undoubtedly, his greatest contribution was when, at the urging of A.J. Muste, he journeyed south to become an advisor to Dr. King during 1956s Montgomery Bus Boycott. He would later write, "I think it's fair to say that Dr. King's view of non-violent tactics was almost non-existent when the boycott began." Dr. King always credited him as a principal influence in developing his views on nonviolence. Rustin's most renowned achievement was as principal organizer of 1963's March on Washington, where his years of practical experience proved invaluable in the complex arrangements of that event. Opponents of the March like the FBI tried to discredit Rustin and sever him from the event by holding his sexuality against him; Rustin was openly gay throughout his life, having once been arrested on a "morals" charge. It is a credit to the movement's leaders that they did not submit to such pressure.

Rustin was typical of the movement's leaders in regarding the championing of nonviolence as a special calling of the African American community. In a 1942 essay, "The Negro and Nonviolence," he wrote that "it is our high responsibility to indicate that the Negro can attain progress only if he uses, in his struggle, nonviolent direct action—a technique consistent with the ends he desires." This was the larger meaning of war resistance.

> Our war resistance is justified only if we see that an adequate alternative to violence is developed. Today, as the Gandhian forces in India face their critical test, we can add to world justice by placing in the hands of 13 million black Americans a workable and Christian technique for the righting of injustice in the solution of conflict.[37]

---

36 D'Emilio, John. *Lost Prophet: The Life and Times of Bayard Rustin*. 2nd ed. University of Chicago Press, 2004; Rustin, Bayard. *Time on Two Crosses: The Collected Writings of Bayard Rustin*. Edited by Devon W. Carbado and Don Weise. 2nd ed. Cleis Press, 2015.

37 Rustin, Bayard. "The Negro and Nonviolence." *PA History*, 1942, https://explorepahistory.com/odocument.php?docId=1-4-184.

Nonviolence has never been just a strategy. It has been a form of politics, meaning it has shaped movements that to one degree or another have conceived themselves as rejecting the models of traditional politics. In recent years, this point has been linked with reclaiming the legacy of Ella Baker, whose formative influence is now acknowledged as equal to any of the above figures. Baker was active in labor movements and civil rights work for two decades before she joined with Bayard Rustin and others in the mid-1950s to organize a group, "In Friendship," to support those engaged in political activism in the South. This led directly to the founding of the Southern Christian Leadership Council (SCLC), of which she was the first Executive Director. She sponsored the gathering that formed the Student Nonviolent Coordinating Committee (SNCC), which she inspired with her vision of organizing: one that eschewed top-down leadership, that focused its efforts not on appeals to elites but on organizing the disenfranchised and the poor, and which practiced a pre-figurative politics that employed methods to build communities of a type that reflected the desired future society. Baker has been acknowledged as an inspiration to the Black Lives Matter movement, whose efforts have been among the most impactful in American history.[38]

[7]/ Pacifism as Dialogue

> In this heroic age, given to war and conquest and violence, the precepts of peace and good will seem to have been almost submerged. The pulpit, the press, and the school unite in teaching patriotism and in proclaiming the glory and beneficence of war; and one may search literature almost in vain for one note of that "Peace on earth, and good will toward men" in which the world still professes to believe; and yet these benign precepts are supposed to be the basis of all the civilization of the western world.
>
> —Clarence Darrow, *Resist Not Evil*

There are a wide variety of practical responses to war described in the above. War abolitionism may be likened to anti-slavery abolitionism in its differing approaches to its goal at any given time and its changing approaches over time. It may even more usefully be compared to movements

38 Ransby, Barbara. *Ella Baker and the Black Freedom Movement: A Radical Democratic Vision*. The University of North Carolina Press, 2003; Parker, Patricia S. *Ella Baker's Catalytic Leadership: A Primer on Community Engagement and Communication for Social Justice*. 1st ed. University of California Press, 2020; Jackson, Sarah J. "Black Lives Matter and the Revitalization of Collective Visionary Leadership." *Leadership*, 17(1), 2021, pp. 8–17.

of anti-sexism or anti-racism today. What we see is not just different approaches but different *kinds* of approach: there are ones that focus on moral suasion of others, or just moral transformation of oneself, versus ones that focus on institutions, or working within existing institutions versus constructing alternative institutions, and so on. I shall identify a larger pattern in pacifist approaches in the next chapter, but my focus now is the variety of perspectives.

One virtue of approaching pacifism as a tradition is illustrating how its different orientations have been in dialogue with each other. The very term "pacifism" was coined at the start of the 20th century to *distinguish* its concerns from the commitments of what was then known as nonresistance. Because of this, some followers of the latter like Mennonites explicitly rejected the pacifist name. The post-World War I developments were partly a response to the perception that aspects of "big tent" pacifism, like calls for mediation institutions, had proved deeply inadequate in forestalling the conflict. Some concluded the causes were deeper, for example, in Western institutions; others concluded their community was elsewhere, specifically women; finally, there was the turn to more expansive approaches to understanding war and more assertive approaches to resisting it.

At the same time, there was significant overlap. In general, no one initially gave much thought to identifying sharp distinctions between the approaches. I have suggested that this is an artifact of later decades, not unrelated to the stigmatizing of the orientation generally. For what is striking is how ecumenical attitudes were and how open-minded pacifists were to differences that were later seen as so monumental. Nowhere is this more evident than in the attitude to Tolstoy.

Tolstoy represented in many respects the most uncompromising form of pacifism, a pacifism that one would think would be dismissed out of hand by anyone concerned with "real-world" change. Hence, just the sort of pacifism that later critics of war might make a point of disavowing.

But this is not how things played out.

It is hardly surprising that Tolstoy and Gandhi maintained friendly connections. Gandhi wrote him in 1909 apprising him of developments in South Africa and raising questions for reflection about the relation of nonresistance to morality. He praised Tolstoy as "the greatest apostle of nonviolence that the present age has produced." Their correspondence continued for over a year, with Tolstoy returning to his spiritual convictions. "[W]hat one calls nonresistance," he wrote in 1910,

> is in reality nothing else but the discipline of love undeformed by false interpretation. Love is the aspiration for communion and solidarity with other souls, and that aspiration always liberates the source of noble activities. That love is the supreme and unique law of human life,

which everyone feels in the depth of one's soul. We find it manifested most clearly in the soul of the infants. Man feels it so long as he is not blinded by the false doctrines of the world. That law of love has been promulgated by all the philosophies—Indian, Chinese, Hebrew, Greek and Roman.[39]

Nor may it be surprising that philosopher William James was attracted to Tolstoy's views. Here is a remark from James's *Varieties of Religious Experience* that references him at length:

No one who is not willing to try charity, to try non-resistance as the saint is always willing, can tell whether these methods will or will not succeed. When they do succeed, they are far more powerfully successful than force or worldly prudence. Force destroys enemies; and the best that can be said of prudence is that it keeps what we already have in safety. But non-resistance, when successful, turns enemies into friends; and charity regenerates its objects. These saintly methods are, as I said, creative energies; and genuine saints find in the elevated excitement with which their faith endows them an authority and impressiveness which makes them irresistible in situations where men of shallower nature cannot get on at all without the use of worldly prudence. This practical proof that worldly wisdom may be safely transcended is the saint's magic gift to mankind. Not only does his vision of a better world console us for the generally prevailing prose and barrenness; but even when on the whole we have to confess him ill adapted, he makes some converts, and the environment gets better for his ministry. He is an effective ferment of goodness, a slow transmuter of the earthly into a more heavenly order.

This is an extraordinarily powerful, and I think moving expression from James (the pragmatist) about the kind of impact that nonresistance can have as practiced by the "saint"—like Tolstoy. The claim is not that such methods will invariably succeed but about the depth of their success when they do; this is, in James's words, the "saints magic gift to mankind": that change is possible which is truly creative, and its very creativity makes it a gesture of hope.[40]

But consider Tolstoy's relation with Jane Addams. She read Tolstoy in school, and her religious views were deeply influenced by him. In the 1890s, along

---

39 Gandhi, Mahatma. *The Collected Works of Mahatma Gandhi.* Vol. 10. Publication Division, Government of India, 1965, appendix VI (ii), pp. 512–514.
40 James, William. *Varieties of Religious Experience, A Study in Human Nature.* Cross-Reach Publications, 2017, p. 116.

with Russian refugees at her settlement house, she studied Tolstoy's theory of nonresistance. She would later describe Tolstoy's book, *What Then Must We Do?* as "a book that changed my life." She devoted a whole chapter to him in her most prominent work, *Twenty Years in Hull House*. She recounts how she too pilgrimaged to Tolstoy's Russian farm where he chastised her for dressing too nicely. Such was his regard for her, though, that he dedicated the profits from his last novel, *Resurrection*, to Hull House.

Most telling, though, was the relationship between Tolstoy and William Jennings Bryan.

The aforementioned William Jennings Bryan was America's dominant progressive political figure. He is the only person to be nominated for president three times by major political party (the Democrats), the only presidential candidate to base an entire campaign on anti-imperialism (1900), and, after becoming Secretary of State in the Wilson administration, he is one of only two such secretaries to resign from principled opposition to going to war (World War I). During the Cold War, when the pacifist streak in progressivism came to be marginalized, Bryan was stigmatized as a religious reactionary for his role in the Scopes trial. But he is arguably the most pro-peace major political figure in American history.

This archetypal "big tent" pacifist regarded Leo Tolstoy as his hero, and the feelings were mutual. In its first year of publication, Bryan's newspaper *The Commoner* praised Tolstoy's opposition to all violence. In turn, Tolstoy was drawn to Bryan after the Spanish-American War and commended his opposition to imperialism in an open letter of 1899. Bryan made a pilgrimage to Tolstoy's Russian estate in 1903 resulting in a 12-hour uninterrupted conversation. Despite his anarchism, Tolstoy endorsed Bryan for president in 1908. When Tolstoy was dying, the only image on his bedroom wall was that of William Jennings Bryan. To the end of his life, Bryan listed Tolstoy behind only the Bible and the speeches of Thomas Jefferson in the books that influenced him.[41]

[II]/ Whatever Happened to Pacifism?

How did pacifism come to be not just marginalized but maligned?

[1]/ Pacifism as "Pathology"

Part of the story involves developments in American progressive political movements in the late 1960s and 1970s. In civil rights, nonviolence

---

41 See Kazin, Michael. *A Godly Hero: The Life of William Jennings Bryan*. Random House, 2006; see also Curti, Merle. *Bryan and World Peace*. Garland, 1972.

generally came to be lumped together with other commitments that younger activists rejected. Critics of Dr. King, Rustin, and others equated their commitment to nonviolence with their belief in integration, multiracial organizations, and so on—which the politics of "Black Power" saw itself rejecting. The issues here were substantial, though it is unclear how much they bore on nonviolence *per se*. Ironically, nonviolence was criticized mainly for its failures as a strategy, but the organizations that broke from nonviolence such as SNCC almost immediately unraveled, while the Gandhi/Dr. King legacy would robustly reemerge in the 1980s.

But this was not before pacifism was roundly dismissed by the romanticizing of violence that characterized some left-wing movements in the United States and Europe during those years. Here, as with the civil rights disputes, I think a factor was very real frustration at the slowness of social change. But no credible alternative models emerged. A document from this era that remains influential as a critique of pacifism is Ward Churchill's "Pacifism as Pathology." It is notable for recycling almost every stereotype and half-truth used to dismiss pacifism in some circles. Pacifism is what unites "white dissident groups," he writes, echoing the claim that "[T]he typical pacifist is quite clearly white and middle class, pacifism as an ideology comes from a privileged context."[42] On the contrary, we have seen that pacifism/nonviolence is the *only* 20th-century political tradition many of whose principal figures have been persons of color/non-Western and women. (Churchill nowhere mentions Cesar Chavez.) His alternative history attributes the major achievements of the American Civil Rights Movement not to nonviolence but to the 1967 rejection of nonviolence by younger political activists. In fact, the major civil rights achievements were years earlier, with little if anything achieved after the rejection of nonviolence. Churchill attributes Indian independence not to the political activities of Gandhi but to Germany's degrading of British forces in World War II. He writes, "their [Gandhi and his allies—CR] victory was contingent upon others physically gutting their opponents for them"—so, Thanks Nazi's!

For Churchill, the politics of Gandhi, Abdul Ghaffar Khan, and Dorothy Day is so bizarre that it requires clinical diagnosis. His explanation: pacifists resist calls to "pick up the gun" due to their "irrational aversion to firearms." Even worse, they gloss over this "deficiency" by "proclaiming it to be a 'moral virtue' and a political dynamic." He claims that this is why progressives wrongheadedly champion such measures as gun control, which is really just a government plot at disempowerment—alongside other nefarious schemes like "antismoking campaigns," which are equally "retrograde and diversionary." What he proposes are workshops on

---

42 Gelderloos, Peter. *How Nonviolence Protects the State*. South End Press, 2007, p. 23.

"demystifying the assault rifle" and a "therapeutic strategy" to include readings from military manuals on the mechanics of booby traps, explosives, demolitions, and so on which he notes are available "at essentially no charge" from the US government.[43] Having dismissed "the pacifist" as invariably white and privileged, he dedicates his essay to the memory of Diana Oughton, a white privileged member of the "Weathermen" whose bomb-making endeavors "might have become the first substantial armed campaign against the state mounted by Euroamerican revolutionaries in the twentieth century" had she not accidentally blown herself up trying to make a bomb.

[2]/ The Pacifist as Moral Idiot

But it is hard to top the derision directed at pacifism from the academic community.

When I entered academia in the 1970s and first thought about writing on pacifism, it became clear that no sin was so great that it could not be ascribed to "the pacifist." Here were some of the allegations:

- Pacifism was responsible for the age's wholesale slaughter of innocents. (Elizabeth Anscombe: "pacifism and the respect for pacifism is [sic] not the only thing that has led to a universal forgetfulness of the law against killing the innocent; but it has had a great share in it.")
- Pacifism was a widespread source corruption, even among those who did not believe it. ("Pacifism has corrupted enormous numbers of people who will not act [?] according to its tenets"—Anscombe again.)
- If pacifists were not cowards, they were still hopelessly confused muddle heads—a bigger sin for philosophers, by the way. (Jan Narveson: "The people who assess conscientious objection as cowardice or worse are taking an understandable step . . . their actions are due, not to cowardice, but to confusion.")
- Or pacifists were just moral freaks. (Tom Regan: "To regard the pacifist's beliefs as bizarre and vaguely ludicrous' is, perhaps, to put it mildly." [One wonders what it would mean to put it *starkly*—CR.])[44]

---

43 Churchill, Ward. *Pacifism as Pathology*. Arbeiter Ring Publishing, 1998. See also Orosco, José-Antonio. "Pacifism as Pathology." *The Routledge Handbook of Pacifism and Non-violence*. Edited by Andrew Fiala, 1st ed., Routledge, 2020.

44 Anscombe, G.E.M. "War and Murder." *Moral Problems*. Edited by James Rachels. Harper & Row, 1971, pp. 269–83 (quotations appear on pp. 279 and 278); Narveson, Jan. "Pacifism: A Philosophical Analysis." *Today's Moral Problems*. Edited by Richard Wasserstrom, MacMillan Publishing Co., 1975, pp. 450–63 (quotation appears on p. 451); Regan, Tom. "A Defense of Pacifism." *Today's Moral Problems*, pp. 464–69 (quotation appears p. 465).

No argument was so idiotic that it could not be ascribed to "the pacifist." Consider this from Martin Caedel, an otherwise astute historian of peace movements: "The pacifist holds that, even if a state fought an extraordinarily successful and costless war, it would nevertheless have committed an impermissible act and would have done better to have submitted stoically to assured butchery and enslavement."[45] Can one imagine any other political position being trivialized in this way? Imagine someone dismissing opponents of sexual trafficking by saying:

> Even if sexual trafficking didn't harm anyone and everyone subjected to it felt happy and fulfilled, and even if it were necessary for the purposes of procreation so that the human race could survive, opponents of sexual trafficking would *still* want to abolish it!

Such dismissals of pacifism are especially striking coming from philosophers, who otherwise evidence not just a tolerance of but an attraction to positions generally deemed unreasonable. Philosophers have had no trouble arguing that "no one knows anything" or that "the world does not exist." The distinguished philosopher Peter Unger once wrote an article titled "I Do Not Exist."[46] But in an era when hundreds of millions of people have been killed by war, the pacifist's unconditional condemnation of it is somehow considered—beyond the pale! Jenny Teichman has noted that these dismissals of "the pacifist" (like all those above) never mention any *actual* pacifist.[47] It is easier to blame pacifists for undermining the law against killing innocents without mentioning Gandhi; it is easier to blame them for widespread corruption without mentioning British pacifist Vera Brittain; it is easier to charge them with being muddleheaded or moral freaks without mentioning Dr. Martin Luther King, Jr.

What is striking, as my historical remarks above have shown, is that pacifism has not always been maligned in this way. Quite the contrary.

The just war tradition arose from taking pacifism seriously. Augustine sought to balance the imperatives of war with Jesus' call to turn the other cheek. The peace voice never entirely disappeared from mainstream Christianity. Erasmus, the most prominent intellectual of the Renaissance, wrote in his *Treatise on War*, one of his most famous writings, that "there

---

45 Ceadel, Martin. *Thinking About Peace and War*. Oxford University Press, 1987, pp. 145–146.
46 Unger, Peter. "I Do Not Exist." *Perception and Identity*. Edited by G. F. MacDonald, Macmillan, 1979. I owe these observations on the contrast with other philosophical positions to Michael O'Connor, see his Oxford BPhil thesis, "Towards a Theory of Pacifism."
47 Teichman, Jeremy. *Pacifism and the Just War*. Basil Blackwell, 1986.

is nothing more unnaturally wicked, more productive of misery, more extensively destructive, more obstinate in mischief, more unworthy of man as formed by nature, much more of man professing Christianity." Pacifism remained a concern, if only as a foil, for early modern just war thinkers. Grotius wrote at the start of *The Law of War and Peace (De Jure Belli ac Pacis)*:

> That war is irreconcilable with all law is a view held not alone by the populace; expressions are often let slip by well-informed and thoughtful men which lend countenance to such a view. Nothing is more common than the assertion of antagonism between law and arms,

he wrote, referencing the early Christian pacifist Tertullian.[48] Vattel began *The Law of Nations or the Principles of Natural Law Applied to the Conduct and Affairs of Nations and Sovereigns* by acknowledging those who take "in a literal sense the moderation recommended in the gospel," while insisting that "The generality of mankind will, of themselves, guard against its contagion."[49] I have already noted the prominence of pacifist ideas at the start of the 20th century.

What changed? How did pacifism go from a respected position to being regarded as the cube of moral idiocy? How did the mark of moral sanity become the willingness to kill people in war?

The main factor, I'd claim, is the rise of total war in the 20th century. Total wars like World Wars I and II were a reversion to the religious wars of the 17th century in which even the slightest form of dissent was demonized. And their claims on the individual were absolute. Universal conscription compelled young men to fight and die for the state. In World War I, the American government sentenced 17 draft resisters to death, 150 to life sentences, and hundreds of others to prion terms ranging from 10 to 20 years (the last draft resistor was not released until 1933).[50] Amidst such a frenzy, pacifists came to be regarded like atheists at a religious revival: their very existence challenged the whole enterprise, so they had to be dismissed as intrinsically immoral or fundamentally irrational, guilty of every sin imaginable.

---

48 Grotius, Hugo. *The Law of War and Peace [De Jure Belli ac Pacis]*. Translated by Francis W. Kelsey, Lonang Institute, 2003, p. 3, www.lonang.com.

49 De Vattel, Emmerich. *The Law of Nations or the Principles of Natural Law Applied to the Conduct and Affairs of Nations and Sovereigns*. Book III ("Of War") Section 3 ("Right of Making War"). Translated by Joseph Chitty, Lonang Institute, 2003.

50 Online at www.encyclopedia.com/history/encyclopedias-almanacs-transcripts-and-maps/draft-resistance-and-evasion

[III]/ Why Pacifism Now?

I referenced some factors in the last chapter involving a larger disillusionment with war and the general defection from participating in it that have made pacifism more tenable. I shall now say some things about recent discussions of just war theory.

[1]/ Walzer's Revival

The Vietnam War regenerated both pacifism and just war thinking. Martin Luther King, Jr.'s 1967 speech "A Time to Break the Silence" was the turning point in antiwar sentiment. King was a "true pacifist" in his words, as were antiwar leaders like A.J. Muste, Dave Dellinger, Barbara Deming, and Dorothy Day. They did not distinguish just from unjust wars; they claimed the illegitimacy of all war. When people of the 1960s generation first heard of "just war" ideals, it was in *response* to such pacifist thinking and in defense of American aggression in Vietnam. Paul Ramsey's writings in the 1960s on "discrimination" and "proportionality," which began the recovery of just war principles, were marshaled in *defense* of American counterinsurgency practices. Ramsey endorsed them "without hesitation." In 1967, he wrote, "No Christian and no moralist should assert that it violates the moral immunity of noncombatants . . . to direct the violence of war upon vast Vietcong strongholds whose destruction unavoidably involves the collateral deaths of a great many civilians."[51]

The pacifist's concern about just war principles was not that they necessarily endorsed crimes like Vietnam but that they were flaccid enough to be employed for any purpose at all.[52] Walter Wink charged that just war thinking was "morally slack." John Howard Yoder claimed that the attraction of the theory lay in its vagueness, making it a useful tool for the powerful. This led to a strange 1973–1974 debate between Gordon Zahn, a prominent pacifist, and James Turner Johnson, the leading historian of just war thinking.

Zahn challenged Johnson to say what—*precisely*—just war principles would say about the Vietnam War, which was well over a decade old by then. Turner began with the striking claim that "there never were very many theologians seriously working within the just war tradition, and now there are even fewer." He continued by saying that "classic just war doctrine as developed within Western Christian thought has comparatively little to say about what may be done in waging war," and concluded that pacifists were

---

51 Ramsey, Paul. "Is Vietnam a Just War?" *Dialog*, 1967, pp. 19–29. Reprinted in Ramsey, Paul. *Just War: Force and Political Responsibility*. Rowman and Littlefield, 2002, p. 53.
52 Wink, Walter. *The Powers That Be: Theology for a New Millennium*. Harmony, 2010.

unreasonable in expecting this (2,000-year-old) tradition to yield practical judgments about an actually existing war.[53] Problems were compounded by the novelty of the Vietnam conflict. It was "not a 'test case' for applying just war standards," as Zahn assumed; it was "a 'test case' for finding out what those standards are, in the sense of what they require in this form of war that took almost everyone in the West by surprise."[54]

Contemporary Anglophone just war theory is a product of the post-Vietnam War era, dating from the publication of Michael Walzer's *Just and Unjust Wars* (1977).[55] While it referenced specific wars of the 20th century, it said nothing about why the 20th century was so warlike or how this might impact our thinking about war generally (in pacifist directions, say). Rather, it stood squarely within the just war tradition in seeking to identify a middle ground between pacifism, on the one hand, and an a-moral "realism," on the other. Insofar as Vietnam had soured people on the whole enterprise of war, Walzer aimed to establish that there could still be "good wars," whose character he sought to define. But in contrast to Grotius and Vattel, Walzer did nothing to engage the views of pacifism despite their prominence in the antiwar movement.

The impact of Walzer's book reflected how little attention there had been to just war thinking, and to war generally, among mainstream political philosophers. Most simply accepted his characterization of the just war tradition, as they still do. Hence, political philosophers have been unappreciative of the full novelty of his efforts. A prominent strain of liberalism had always been highly critical of war. Peter Gay's influential study of the Enlightenment from just a few years earlier dwelled at length on its denunciations of war.[56] This critique was muted in the 19th century by nationalism. It came to be assumed that war *per se* was a massive violation of individual rights but that this could be discounted since the claims of the community trumped them in times of war. This was the view of America's leading liberal thinker in the first half of the 20th century, John Dewey. The waning of nationalism later in the 20th century meant that the tension between war and individual rights reemerged as a problem, as in the Vietnam War debates.

53 Johnson, James Turner. "Rationalizing the Hell of War." *Worldview*, 17(1), 1974, pp. 43–47. in response to Zahn, Gordon. "War and Its Conventions." *Worldview*, 16(7), 1973, pp. 25–33.

54 Johnson, "Rationalizing the Hell of War."

55 Walzer, Michael. *Just and Unjust Wars*. Basic Books, 1977.

56 Gay, Peter, *The Enlightenment: The Science of Freedom*. Revised edition. W. W. Norton & Company, 1996, p. 401, ff. Gittings, John. "The Growth of Peace Consciousness: From the Enlightenment to the Hague." *The Glorious Art of Peace: From the Iliad to Iraq*. Oxford University Press, 2012.

Walzer's approach was novel in the robustness with which it made the case for liberal war. That is, it did not just claim that war was *compatible* with individual rights; it claimed that war was *compelled* by those rights. Thus, *Just and Unjust Wars* began with Walzer's view that "the arguments we make about war are most fully understood . . . as efforts to recognize and respect the rights of individual and associated men and women." His account of just and unjust wars, that is, was grounded in the "doctrine of human rights." This privileging of individual rights fit with the rights-based orientation associated with figures like John Rawls that was becoming hegemonic at the time.

But anyone involved with the political conflicts around the Vietnam War would invariably ask how this insistence on the primacy of individual rights squared with the institution of *conscription*. Starting with Hobbes and Locke, rights-oriented theorists had wondered about the status of soldiers in war: how could the demands for sacrifice on them be reconciled with fundamental rights like that of self-defense? Walzer portrayed his book as provoked by the debates of the Vietnam War, of which the argument over the draft was crucial. But he said nothing about conscription in his book. And everything he said about soldiers implied that their fate had nothing to do with rights; indeed, one of the central arguments of his book (for the "moral equality of soldiers") rested on the *denial* of their rights. Their condition was one of "military servitude" subject to "the thralldom of the trenches." And he acknowledged the larger point:

> "States exist to defend the rights of their members, but it is a difficulty in the theory of war that the collective defense of rights renders them individually problematic. The immediate problem is that the soldiers who do the fighting, though they can rarely be said to have chosen to fight, lose the rights they are supposedly defending. . . . "Soldiers are made to be killed," as Napoleon once said; that is why war is hell."

"Difficulty," indeed. It's hard to see how war can be a "collective defense of rights" if those defending them *lose* those rights—through no fault of their own.

Walzer painted an attractive picture of what war *ought* to be on the liberal model. But what did this have to do with what war actually was? Rousseau chastised just war theorists for confusing the two, for confusing "right" with "fact" in his words. Walzer was mindful of this concern; he claimed that his account was articulating what he termed the "moral *reality* of war" and not some moral *fantasy*. But his argument consistently skirted between the two.

A case in point was his invoking what he termed the "domestic analogy." That analogy, he maintained, was "indispensable" to just war thinking—despite the fact that, as he acknowledged, "international society as it exists today is a radically imperfect structure." Indeed, it was *so* "imperfect" that the analogy hardly held at all. War was "like" domestic punishment, except that "neither the procedures nor the forms of punishment have ever been firmly established in customary or positive international law. Nor are its purposes entirely clear: to exact retribution, to deter other states, to restrain or reform this one?" So, it was like domestic punishment except it lacked all the features that made such punishment legitimate. War was "like" law enforcement, except "there are no policemen," just sovereign states acting on the principle of "self-help," their wars the equivalent of "citizen's arrests"—another name for which might be "taking the law into their own hands." And the "law" was a primitive one: civilized society treated different crimes differently, but international society treated every crime the same. "It is as if we were to brand as murder all attacks on a man's person, all attempts to coerce him, all invasions of his home." We do this, Walzer said, because otherwise the whole structure would collapse. International society "might be likened to a defective building, founded on rights; the whole thing shaky and unstable because it lacks the rivets of authority."

The basic problem here is that if states are "like" persons in some sense, they are equally *unlike* them in being organized collectives—the very dimension that makes their actions *war*. Walzer acknowledged as much, suggesting that because of this their violence was more a "feudal raid," like the Vikings descending on Paris or the Magyars on Northern Italy, and their "self-defense" more like the counter-raids of Otto the First. This was why war even as an act of punishment "will more probably extend than cut off the violence." In summary, then states were like vigilantes, armies like posses, and the "punishment" they inflicted was like lynching for even petty offenses, or shoot-outs, where rights are affirmed by the law of the gun.

Walzer himself admitted at one point that the better analogy might be "the 'wild west'" of American fiction, which is why during the Vietnam War people found analogies to "international society" in films like "Hang 'Em High" or "The Wild Bunch." Or an even better analogy: "The Godfather," the most significant movie to appear in the Vietnam War years, about a "war" as a form of organized crime between Mafia families that were nothing more than protection rackets.

[2]/ Revisionism

Walzer took the Realist critique of just war thinking seriously and ignored pacifism. This remains true of much international relations theory. By contrast, Walzer's Revisionist critics demote Realism and tend to take the pacifist challenge to just war thinking seriously.[57] This has led to a serious and sustained interest in pacifism and nonviolence of a type that is unprecedented in Anglophone political philosophy.[58]

Revisionists have accepted much of Walzer's framework. They have adopted his rights-oriented framework. They have accepted his account of the "traditional" just war framework, with little concern for the tensions within it and how it has changed over time. Finally, they have adopted a generally ahistorical/anecdotal approach to war, with little concern for war's full institutional dimensions and how it has dramatically changed over time. All of this contributes to Revisionism's "abstraction" which critics have found worrisome.

A case in point is the discussion of soldiers, a central concern of Revisionism. There is little concern for how the institution of soldiering has dramatically changed from era to era; in particular, while attention has been given to what soldiers *do* there has been little attention to who soldiers *are*: what are the ways in which societies have mobilized its members for soldiering, which has changed dramatically. This raises an issue that was first raised by Clausewitz, and is addressed in the next chapter: the penchant to think of war solely as what happens on the battlefield, and ignore what happens before/beyond the battlefield. A further point involves the focus on "soldiers" to the exclusion of other types of combatants and the issues they raise. The dominant empires have tended to be *maritime* ones. This was certainly true of the British Empire, the hegemonic military force of the 19th century. Theirs was not a story of soldiers and armies but of sailors and navies—where the issues raised by conflict assumed a quite different character.

But Revisionists depart from Walzer in key respects, some of which include the following:

- They demand a philosophical *precision* which Walzer's more casual approach did not seek. In so doing, they implicitly respond to the aforementioned pacifist concern that just war principles are overly vague.

---

57 Lazar, Seth. "War." Edited by Edward N. Zalta. Stanford Encyclopedia of Philosophy, 2016. https://plato.stanford.edu/entries/war; Rengger, Nicholas. "On the Just War Tradition in the Twenty-First Century." *International Affairs*, 78(2), 2002, pp. 353–363; 354. For an overview, see Lazar, Seth. "Just War Theory: Revisionists vs. Traditionalists." *Annual Review of Political Science*, 20, 2017, pp. 37–54.

58 See the recently founded *Journal of Pacifism and Nonviolence*. For an overview of current developments, see Christoyannopoulos, Alexandre. "Pacifism and Nonviolence: Discerning the Contours of an Emerging Multidisciplinary Research Agenda." *Journal of Pacifism and Nonviolence*, 1(1), 2023.

- In so doing, they have revived previous liberal concerns about an inherent *tension* between war and individual rights.

This reflects a more stringent conception of individual rights, developed with a sophistication and thoroughness unprecedented in the discussion of war. Special attention has been given to the principles of *jus in bello* as grounded in such rights. Can war be fought in ways that respect those principles? For example, should anyone take the moral risk of fighting in a war if the chances of violating those principles are so great? These are the sort of questions that have given plausibility to forms of contingent pacifism.

- They are more given to employing *hypothetical* examples. This fits with their more analytical approach generally.
- Most importantly, I think, Revisionists have been more willing to follow the philosophical arguments where they lead. In contrast to so much past discussions, they do not regard pacifist-type conclusions as *prima facie* grounds for throwing out an argument entirely.

Walzer's approach was what Thorstein Veblen once termed, in critiquing conservative economics, the principle of *ceremonial adequacy*. He meant by this the principle that an argument ultimately stands or falls on whether it confirms what is already believed. In matters of war, this obviously discounts pacifism from the start. As Revisionists note, it discounts any significant departure in thinking about war from the *status quo*.

Does this put things too starkly?

Walzer's *Just and Unjust Wars* actually garnered little attention when it appeared. There were a number of reasons for this. People were sick of war generally in the immediate post-Vietnam years. The two issues that came to dominate war discussions in the 1980s, nuclear weapons and American interventions in Central America, were ones that Walzer had little to say about. As Walzer himself has noted, *Just and Unjust Wars*—and just war thinking generally—"came into their own" only in the 1990s with the first Gulf War. It was a war explicitly designed to re-legitimize war against what was termed the "Vietnam Syndrome." Walzer's insistence that there could be just—liberal—wars fit with the emerging perspective.

Naysayers remained, of course. This was still a time of disillusionment with and defection from war. Walzer addressed this in a telling article of 2002, "The Triumph of Just War Theory (and the Dangers of Success)." He put it in terms of the different audiences.[59] The diehard peaceniks of the

---

59 Walzer, Michael. "The Triumph of Just War Theory (and the Dangers of Success)." *Social Research*, 69(4), 2002, pp. 925–944.

Vietnam Era championed what he termed "doctrines of radical suspicion" about war because theirs was "a radicalism of people who do not expect to exercise power." "By contrast, just war theory, even when it demands a strong critique of particular acts of war, is the doctrine of people who do expect to exercise power."

Just war theory, thus construed, is a policy exercise—specifically, one whose animating purpose is to advise those who "exercise power." This could include *some* criticism of what "those in power" do, especially in *jus in bello* matters; but it obviously could not depart too much from the thinking of "those in power" if it was to maintain its purpose. So its animating impulse could not be that of Revisionist thinking of following arguments where they lead and worrying later about what "those in power" think. Walzer's position is a strange one given his record of activism against the Vietnam War. Many of the debates of that war among ordinary people, including soldiers, were not conducted in terms congenial to "those in power," indeed some of them were ones of "radical suspicion"—yet they were politically impactful nevertheless.

[3]/ Toward a Critical War Theory

> One of the difficulties we face in calling for the abolition of war is whether we know what we are calling to be abolished. War may be ubiquitous, but that there always seems to be a war taking place somewhere does not mean that we know what we mean when we say war. Clearly war is a contested concept that requires analogical display.
>
> —Stanley Hauerwas[60]

Yet the Revisionist framework is too narrow to resolve the question: Pacifism, pro or con? One reason has been noted already. Revisionism, like traditional just war thinking, focuses on the appraisal of individual wars, whereas pacifism as war abolitionism rests on an appraisal of the war system, and such an appraisal cannot proceed via the ahistorical/anecdotal/hypothetical approaches adopted. It requires a much more expansive approach to the subject matter.

Duane Cady has written, "[T]he war system, the standard operating procedure of sovereign states constantly preparing for, threatening, and employing military force in domestic and international affairs, goes almost

---

60   Hauerwas, Stanley and Hogan, Linda and McDonagh, Enda. "The Case for Abolition of War in the Twenty-First Century." *Journal of the Society of Christian Ethics*, 25(2), 2005, pp. 17–35.

wholly unquestioned."[61] One reason the war system goes unquestioned is the simple fact that the topic of war generally goes unaddressed. Remarkably, I do not know of a single extended discussion of what—precisely—"war" is in political philosophy. As Hauerwas suggests, arguments about war just presume what "war" is ("the sort of thing that happened in World War II") and then zip along to the fine details.

The blindness to the whole topic of war has been recognized in other fields for some decades now. Fernand Braudel wrote, "[W]e are as ignorant of war as the physicist is of the true nature of matter."[62] Anthony Giddens has written of social theory in general,

> I have long contended that the neglect of what any casual survey of history shows to be an overwhelmingly obvious and chronic trait of human affairs—the resort to war—is one of the most extraordinary blank spots in social theory in the 20th century.[63]

Martin Shaw writes,

> In the history of 20th-century social theory, however, war hardly figures. Our most important writers have continued the debates of 19th century thinkers about industrialism and capitalism, recognizing war as an event external to the main prosthesis of social change—if at all. The late 20th century boom in sociology, even radical and Marxist, has uncritically taken 1945 as the baseline of modernity, failing to reflect on the processes of war, which determined this major rupture in social history.[64]

Michael Mann, Charles Tilly, Theda Skocpol, and others have stressed the same point. Like Shaw, Mann attributes it to sociology's continued reliance on 19th-century models of society fashioned before the era of total war.[65]

61 Cady, Duane. *Warism to Pacifism: A Moral Continuum*. 2nd ed. Temple University Press, 2010, p. 21.

62 Cited in Manicas, Peter T. *War and Democracy*. Blackwell Publishing, 1989, p. 1.

63 Cited in Shaw, Martin. *Global Society and International Relations: Sociological Concepts and Political Perspectives*. Cambridge: Polity Press, 1994, p. 38.

64 Shaw, Martin. "War and the Nation State in Social Theory." *Social Theory of Modern Societies: Anthony Giddens and His Critics*. Edited by David Held and John Thompson. Cambridge University, 1989, p. 129.

65 Mann, Michael. "War and Social Theory." *The Sociology of War and Peace*. Edited by Colin Creighton and Martin Shaw, McMillan, 1987.

Hans Joas notes in his *War and Modernity* that the relation of war to social life generally has been ignored.[66]

What of political philosophy?

I have suggested that no one reading mainstream political philosophy would have any idea that the 20th century occurred at all—as a century of unprecedented violence. This is quite remarkable given the context in which it emerged. The writings of sociologists, especially British ones, were deeply impacted by political events such as the anti-nuclear campaigns of the 1950s–1960s. By contrast, the "revival" of political philosophy inaugurated by John Rawls in the 1970s occurred against the background of the Cold War and in the midst of the conflict surrounding the Vietnam War. But it showed basically zero interest in the question of war. Rawls is often portrayed as responding to the political conflicts of the 1960s, but this is only half-true. His was a response to the Civil Rights Movement, but it ignored the anti-Vietnam War Movement and the questions raised for example of how the greatest threats to civil liberties arose from the claims of the national security state. Accordingly, war became invisible as a topic. It is not mentioned in the major introductions to political philosophy, such as Will Kymlicka's *Contemporary Political Philosophy: An Introduction* or Jonathan Wolff's *An Introduction to Political Philosophy;* the term "war" does not appear in the index of Jurgen Habermas' two-volume *Theory of Communicative Action* and so on. Postmodern discussions are generally useless, dwelling on "subtle" forms of covert violence and the machinations of "micropower"—while ignoring the gross forms of power and violence that is war.[67] Jonathan Glover has remarked on how this impacts philosophy tone. "Even in applied ethics awareness [of the atrocities of the 20th century—CR] is often missing. The tone of much writing suggests that John Stuart Mill is still alive and that none of the 20th century has happened."[68]

Where does this leave us in confronting the challenges of the 21st century?

In his *Century of War: Politics, Conflicts, and Society Since 1914* (1995), Gabriel Kolko remarked on this blindness in the case of historians. He wrote:

In both Europe and Asia, wars have been the single most important catalyst of the great political and ideological transformations that have

66 Joas, Hans. *War and Modernity: Studies in the History of Violence in the 20th Century.* Translated by Rodney Livingstone, Polity, 2003.

67 The exception is Michel Foucault's *Society Must Be Defended: Lectures at the College De France*, 1975–76. Picador, 2003.

68 Glover, *Op. cit.*, pp. 5–6.

characterized [the 20th-century], and they remain humanity's principal nemesis both today and in the future.

Indeed,

> It is astonishing that at the end of nearly a century of monumental and increasingly unrelenting conflict that virtually all of the influential currents in social theory should, each in their own fashion, essentially marginalize contemporary history's central political, social, and military experiences.

And Kolko strikes a deeply cautionary note about what this blindness portends for the future:

> At the end of this century we simply do not know nearly enough—nor have we thought sufficiently—about the social and human processes and events that have gone into the making of our century, and that in many decisive ways have made the future of humanity more precarious than it has been at any time in the past 50 years.[69]

Political philosophy needs what I term a *critical war theory*. Such a theory does not presume a commitment to pacifism. On the contrary, its framework allows for *posing* the question "Pacifism, pro or con?" in ways that more constricted approaches do not. I hope that some of its features are exemplified in the current work, but let me state a few of them:

- Currently, "just war theory" is rather incidental to the concerns of political philosophy generally. There is a historical background to this. In the 19th century, British liberal thinking—in contrast to Continental thinking—came to see war as incidental to the polity's principal concerns. This was not unrelated, I would argue, to the marginalization of empire in British political/social theory at exactly the same time that empire became the dominant fact for British life.

In any event, critical war theory blurs the distinction between them by reasserting war as a central concern of the state, such that no interesting questions about the state can be addressed by abstracting from the question of war.

- Critical war theory insists that war must be approached historically and institutionally, in much the same way that political philosophies of

69 Kolko, Gabriel. *Century of War: Politics, Conflicts, and Society since 1914.* New Press, 1994, p. x; 464.

gender and race now insist for their subject matters. So the division between political philosophy and the sociological/historical study of these matters must also break down.

- The story of war has not just been a story of states/empires/nations mobilizing for war; it has also been one of movements organizing against war. The role of social movements has been increasingly appreciated in writings on racism. This reflects the fact that any attention to institutions must involve attention to institutional change, which itself must consider social movements. So a critical war theory must attend to the movements against war, addressed here under the umbrella of pacifism.

Chapter 3

Personal and Political Pacifism

This chapter provides a framework for understanding the pacifist tradition by exploring two distinctions that are crucial to any philosophy of pacifism. The first is the distinction between the two basic concerns of pacifism that I call *appraising* and *opposing*. This is roughly the theory/practice distinction, so it is one we encounter in other practically oriented approaches. The other is the distinction between two different kinds of pacifism that I call *personal pacifism* and *political pacifism*. The previous chapter stressed the variety and complexity of the pacifist tradition. This distinction marks the most significant difference in pacifist perspectives. They are not mutually exclusive; on the contrary, major figures like Gandhi and Dr. King have adopted both. The discussion of political pacifism introduces the notion of the war system in its basic elements and the critique of that system in its various dimensions. These are further developed in the next chapter. I conclude with some remarks on the problem of "absolutism" and a defense of the uncompromising form of pacifism endorsed in this work.

[I]/ The Two Concerns of Pacifism: Appraising and Opposing

The two basic concerns of pacifism are important to distinguish if only because arguments against pacifism typically mush them together. Let me introduce them by considering another radical movement, the 19th-century abolitionist movement against slavery in America.[1]

At the time, "abolitionism" denoted two things. An abolitionist was someone whose views of slavery were ones of unconditional condemnation.

---

1 On the different approaches to abolition, see Kraditor, Aileen S. *Means and Ends in American Abolitionism: Garrison and His Critics on Strategy and Tactics 1834–1850*. Elephant Paperback ed., 1989.

DOI: 10.4324/9781032686189-3

They did not distinguish good and bad instances of slavery; they condemned it without *qualification*. And an abolitionist was someone whose actions toward slavery were ones of unconditional opposition. They opposed it, as it were, without *reservation*. One without the other was not enough. There were people whose attitudes to slavery were ones of unconditional condemnation, but their reasons were such that they implied nothing could be done about it; hence, they were not considered abolitionists.

Accordingly, I take pacifism to denote two things. On the one hand, it is a theoretical perspective that is a way of *appraising* war. As I construe it, its appraisal is one of *unconditional condemnation*. Pacifism does not distinguish good and bad wars; it condemns war without *qualification*. On the other hand, it is a practical position that is a way of *opposing* war. As I construe it, its practical position is one of *unconditional opposition*. Pacifism opposes war without *reservation*.

Thus construed, pacifism has faced two different objections. Its unconditional condemnation of war invites the charge of *dogmatism*, and its unconditional opposition to war invites the charge of *fanaticism*. Note that this is an objection to the *kind* of position pacifism is, often put in terms of its being an "absolutist" one. I shall prefer the term uncompromising. As such, it pertains to other positions of this sort. For example, 19th-century abolitionists were also accused of being dogmatists and fanatics. Some regard the uncompromising opposition of critics today to practices like torture as dogmatic.

The charge of dogmatism/fanaticism is of special concern to the pacifist for this reason.

Many have seen a connection between "absolutist"-type thinking and violence. They have held that the one invariably leads to the other. Gandhi once asked how anyone who thinks they possess "absolute truth" could be "fraternal." Other pacifists have raised the same concern. I think it is partly in response to this that some have recently proposed less "absolutist" forms of pacifism such as versions of "contingent" pacifism. I return to this at the end of this chapter.

What needs stressing from the start is that even within pacifism so construed there is a great deal of room for disagreement, as evidenced in the overview of the previous chapter. Indeed, it is hard to find any point on which there has been complete agreement on *why* war should be condemned or *how* should be opposed—beyond the fact that it should be condemned and opposed. But one disagreement stands out. With abolitionism, a major disagreement was what *counted* as "slavery." A case in point was *prisons*: some felt that prisons were also slavery; others did not. This debate continues, as does the debate about whether, say, egregious forms of labor exploitation constitute slavery. In any event, it was to avoid this debate that the 13th Amendment to the United States Constitution specifically exempted prisons from slavery.[2] Similarly, pacifists have disagreed about what *counts* as "war." Most notably, 19th-century pacifists disagreed on whether collective self-defense counted as war.

To contemporary eyes, this may seem like fudging given the prominence—or more accurately, inflation—of self-defense in today's justifications of war. (To say "I don't believe in war, but I do believe in collective self-defense" seems inconsistent.) But prior to the 20th century, self-defense was often distinguished from "war" proper. For example, the United States Constitution can be seen as distinguishing them.[3] More recently, pacifists have disagreed whether *collective security* actions count as war. The thought here is that political pacifists generally do not oppose domestic law enforcement, so need not oppose law enforcement if it could be achieved on the global level. These have significant practical consequences. Distinguishing self-defense from war leads to proposals for self-defense that do not involve the evils that war normally involves; the same holds for proposals for collective security. Ultimately, these matters not be resolved in the abstract. What counts as "war" and what constitutes its evils involves critical attention to the practice itself.

Arguments *both* pro and con pacifism often mush the appraising/opposing distinction together. One of the most common objections to pacifism rests on this. "Pacifism" derives from the term for peace, but it is often misconstrued as meaning "passive"—and then it is denounced for acquiescing to evil. To begin with, none of pacifism's practical approaches is "passive" in this sense. But prior to this, the objection rests on equating what is both a theoretical and a practical position with the latter alone. This is conceptually confusing and contributes to bogus dismissals of it. For example, critics of pacifism note that some pacifists have adopted rather foolish ways of opposing war and then conclude that their condemnation of war must be false. But *any* position, certainly any radical position, can be erroneously dismissed on such grounds. Some 19th-century abolitionist opponents of slavery engaged in foolish or ill-considered actions to end slavery. Many regarded John Brown's violent actions at Harper's Ferry to be deeply ill-conceived. But it was mistaken to conclude from this that the abolitionist critique of slavery was false.

---

2 It reads, "Neither slavery nor involuntary servitude, except as a punishment for crime whereof the party shall have been duly convicted, shall exist within the United States, or any place subject to their jurisdiction."

3 For example, the founders of the United States identified war with an act of the federal government, requiring a formal declaration; self-defense was identified with local militias who needed no such declaration.

A fundamental issue for the philosophy of pacifism is the relation *between* appraising and opposing. This is basically the question of theory and practice; so it is one that other practically oriented political perspectives also face. But the question is especially significant for radical positions like pacifism, or earlier abolitionism. Their appraisal is so negative, and all-encompassing that it might seem that *any* practical opposition is insufficient to the task; indeed, it can seem outright silly. The pacifist's indictment of war can be so all-encompassing that some may conclude it is hopeless to do anything about it. Note that the issue can go in the other direction, from opposing to appraising. This is raised by the very great prestige of nonviolent practices today, which some have contrasted with the continuing marginalization of pacifism.[4] One might ask: what appraisal of the war system is implicit in the various nonviolent strategies that people have developed for opposing violence? Does nonviolence, as a form of opposing, presume pacifism, as a form of appraising?

[II]/ The Two Types of Pacifism: Personal Pacifism

Personal pacifism, as I construe it, condemns killing as an individual act. Because it condemns killing as an individual act, personal pacifism condemns any social practice involving that act. This is why it condemns war, as well as practices like capital punishment. Its condemnation of killing also implies a rejection of personal self-defense, as we saw in remarks like those of Garrison and Tolstoy. But more must be said about the nature of that "rejection," since it has not implied a *condemnation* of those who kill in personal self-defense.

This bears on the question of *why* personal pacifism condemns killing as an individual act. Some personal pacifists condemn all killing because they condemn all violence or force; hence, condemn all killing as instances of violence or force. This is how proponents of "Nonresistance" have often put it. It obviously raises questions of how to pin down notions like "violence" or "force," which is why I prefer construing it as opposition to killing. But linking it to notions like "force" is illuminating in this respect. Personal pacifism has almost always condemned the taking of human life as an illicit use of *power*—as claiming, for example, an authority over life and death that properly belongs to God alone. It is a form of self-idolatry. This is crucial for understanding the pacifism of the first Christians whose rejection of killing involved identifying it with the illicit power over life and death claimed by Rome and its Emperor. Early modern Christian pacifists,

---

4 Howes, D. E. "The Failure of Pacifism and the Success of Nonviolence." *Perspectives on Politics*, 11(2), 2013, pp. 427–446.

responding to the rising market order, equated this with treating human life as a commodity instead of a gift, as something that could be "taken" like any other piece of property.

The origins of this pacifism are invariably religious. It is hard to find anyone in the Western tradition who does not trace their pacifism of this sort to first Christianity, specifically the teachings of the Gospel. Some of its proponents regard it as a specifically sectarian position, that is, one that is only intelligible in the context of the Christian witness. I take this to be the view of John Howard Yoder and Stanley Hauerwas, the foremost proponents of such pacifism today.[5] But I do not find this in the views of Garrison or even Tolstoy, despite their strong references to Christianity. What I do find is an emphasis on the *spiritual*, and the suggestion that pacifism of this sort is evoked in certain types of experiences. I say more about this in a moment.

While such pacifism need not be sectarian, in my view, it is *parochial* in this sense. Its condemnation of killing does not claim to be an "ethic for everyone." This is most evident in the issue of self-defense. Its condemnation of all killing has meant an absolute refusal to engage in such killing oneself, even in self-defense, but this has not extended to a condemnation of others who do so. This does not mean that its ethics is just a matter of personal preference. Killing by others even in self-defense can be appraised, but the terms of that appraisal are not those of "condemnation"—or even those of judgment as they would be if the ethic were "universalizable" in the standard sense. This bears on the concern with "absolutism," voiced above. For me, what is intriguing about personal pacifism is partly how its ethic does not fit into our standard conceptions.

Implicit in its condemnation of war is that killing is a constitutive part of war and not a purely contingent one. I mention this only because some have portrayed the absurdity of pacifism as such that it would still condemn war even if it involved no killing. There can of course be disagreements about the constitutive versus contingent features of a practice. But if anything is constitutive of war, killing is; a war without killing is a "war" only metaphorically. But note then that this is not the only form that personal pacifism might take. For example, someone might with good reason regard *rape* as a constitutive part of war. It has certainly been a persistent

---

5 See, for example, Yoder, John Howard. *The War of the Lamb: The Ethics of Nonviolence and Peacemaking.* Brazos Press, 2009; Yoder, John Howard. *Nonviolence—A Brief History: The Warsaw Lectures.* Baylor University Press, 2010; Hauerwas, Stanley. *The Peaceable Kingdom: A Primer in Christian Ethics.* University of Notre Dame Press, 1991; Hauerwas, Stanley. *War and the American Difference: Theological Reflections on Violence and National Identity.* Baker Academic, 2011.

feature of war as we have known it, and some may hold that it better reflects the essentially gendered nature of war. So, someone might *not* condemn all killing but still condemn all war because they condemn all rape, and hence condemn all war as involving that act.

[1]/ Personal Pacifism's Appraisal of Killing

How should we understand personal pacifism's condemnation of all killing?

My principal focus in this work is political pacifism. But I want to say some words about personal pacifism partly because its perspective on killing merits serious attention and partly because the perspective of political pacifism is illuminated by the contrast.

Personal pacifism has no single ethic. About the only generalization one could make is that its perspective is distinct from the juridical one of "rights" and "duties" that dominates today's political philosophy and defines its just war thinking. Most of its strongest voices today are in fact religious ones, some of whose perspectives are unabashedly grounded in their religious commitments. As I read Yoder and Hauerwas, their claim is that pacifism follows from a full acceptance of Jesus' ministry and message. Being a Christian means following the example of Jesus beginning with the call to turn the other cheek, whose meaning is itself defined by Jesus' own actions. This is not a purely personal ethic. It is realized in the institution of the church which stands as a critical alternative to the violence of the political powers and as a prophetic anticipation of the Gospel's fulfillment. The reasoning proceeds from what it *means* to be a Christian, which is just to say that it does not take itself to be speaking to everyone. But this is not a shortcoming, since for this perspective, there is no ethic that speaks to everyone; indeed, those claiming to speak to everyone prove to be exercises of accommodation to the *status quo*. Rather, ethics always proceeds from within a particular tradition—in this case Christianity.

By contrast, let me offer a rendering of the personal pacifist ethic that is faithful to the words and practices of some of its key figures and movements but does not presume any particular religious perspective. I stress that this is just my own interpretation of it, and a sketch at that. It will succeed if it inspires others to inquire further into this approach.[6]

---

6 For further discussions of this ethic, see Ryan, Cheyney. "The Morning Stars Will Sing Together: Compassion, Nonviolence, and the Revolution of the Heart." *The Routledge Handbook of Love in Philosophy.* Edited by Adrienne Martin, Routledge, 2019; and Ryan, Cheyney. "The Lament of the Demobilized." *To End a War: Essays on Justice, Peace, and Repair.* Edited by Graham Parsons and Mark Wilson, Cambridge University Press, 2022.

Personal pacifism in the West originated with the first Christians, whose rejection of war arose from a rejection of Rome's imperial arrangements and their warlike proclivities. This critique of empire was fundamental, as it has been for all pacifisms. All killing was wrong, but killing in war was especially wrong because it was *idolatrous* in ceding a power to the state over life and death that properly belonged to God alone. This critique of idolatry has persisted in all forms of personal pacifism, or thinkers deeply committed to nonviolence. A recent example is Emmanuel Levinas, whose critique of violence is cast in terms of treating the Other as just another thing by responding to the Other's face as a "mask"—something "fixed" and "frozen," like an idol; all in the name of claiming an illicit power over them. In the same vein, Martin Buber speaks of turning a "Thou" into an "It."[7] This suggests that, while all personal pacifisms have not presumed a particular religious perspective, they have all had a religious inflection. This helps explain why it has often moved people in ways that political pacifism's juridical perspective has not.

I take this religious inflection to mean two things.

One is its orientation to the *sacredness of life*. From the juridical perspective, the focal point is the "person," a category that the early rights theorists adopted from Roman law: illicit killing involves a violation of the person's rights, specifically the right to life. For personal pacifism, the focal point is life and killing involves a violation of its sacredness. How should we understand that "sacredness"? A starting point is the rendering of killing as "taking" human life. Talk of "taking" evokes the picture of killing as a type of stealing, that killing another means taking something that is rightfully their possession; as if life were one's private property. But then it is hard to make sense of talk of taking one's *own* life. A better picture, and one in keeping with the larger perspective of this tradition, is of life as a *gift*. The gift of life inscribes us in a larger cycle of giving and receiving that Dr. Martin Luther King, Jr. described as a "web of mutuality." The "taking" of life is best construed then as the removing of life from that web of mutuality. This would explain the parallels between killing someone and enslaving them. The latter imposes a kind of social death, removing persons from the cycle of giving and receiving/mutuality of which they are properly a part, as children, spouses, and ancestors.

How do we come to appreciate that sacredness? This brings us to the second religious inflection, the appeal to *experience*.

---

7 Levinas, Emmanuel. *Totality and Infinity*. Translated by A. Lingis, Duquesne University Press, 1969; Buber, Martin. *I and Thou*. Translated by Ronald Smith, Charles Scribner's Sons, 1958.

The juridical rights-oriented framework approaches things from an impersonal third-person perspective. By contrast, personal pacifism's approach is an interpersonal, second-person orientation. For it, the sacredness of human life is not something that is established by argument but rather something that is revealed to us in experience. There is a famous passage in *Anna Karenina* where Tolstoy describes how Levin—the author's self-projection—is awakened to the sacredness of life by the witnessing of birth.[8] This evokes the notion that sacredness is especially associated with the mysteries of life, as marked by such moments of deep transition as birth, marriage, and death. It finds its place in the fact that life is itself a journey, such that appreciating its sacred moments involves situating ourselves between the past and the future.

For personal pacifism, the sacredness of life can also be revealed in the *resistance* to taking life. This resistance is often portrayed in terms of the challenge of killing another when one fully confronts the face, often specifically the eyes, of the victim. In the famous antiwar novel, *All Quiet on the Western Front*, Erich Remarque describes how the eyes of the enemy, in his words, "cry out," and "yell" in ways that are disempowering.

> But you were only an idea to me before, an abstraction that lived in my mind and called forth its appropriate response. It was that abstraction I stabbed. But now, for the first time, I see you are a man like me. I thought of your hand-grenades, of your bayonet, of your rifle; now I see your wife and *your face* and our fellowship. Forgive me, comrade. We always see it too late. Forgive me, comrade; how could you be my enemy?[9]

Or take the case of the former Japanese Zero fighter, Kaname Harada, whose wartime experiences led him to pacifism and committed antiwar activism. He described how when close to his opponents in dogfights he saw "their terror-stricken faces before gunning them and their aircraft down to their deaths." In his horrible dreams, he said, he could "see the terrified faces of the American pilots he had shot down during the war." "That is how war robs you of your humanity by putting you in a situation where you must either kill perfect strangers or be killed by them," he stated.[10]

8 Tolstoy, Leo. *Anna Karenina*. Translated by Rosemary Edmunds, Penguin, 1978, p. 749. I was drawn to this discussion by the discussion in Ronald Dworkin's *Life's Dominion*. Knopf, 1993, p. 83.
9 Remarque, Erich Maria. *All Quiet on the Western Front*. Ballantine Books, 1987, p. 223.
10 Wallace, Rick. "WWII Japanese Pilot's Quest for Redemption." *The Australian*. April 3, 2015; Roberts, Sam. "Kaname Harada, Pearl Harbor Fighter Pilot and, Later, Remorseful Pacifist, Dies at 99." *The New York Times*, May 5, 2016.

Anglophone ethical theory, I would note, is not entirely unfamiliar with such talk. Thomas Nagel grounds one of his accounts of strong moral prohibitions with talk of how certain actions strike us as "swimming against the moral current." He also notes that accounting for such resistance is itself a challenging task, to the point that it is not easily done in straightforward impersonal talk.[11] The juridical rights-oriented framework understands the prohibition on killing as a matter of duty, as involving what we "ought not" do. By contrast, personal pacifism construes the sacredness of life as involving what we "cannot" do. This is a category explored by contemporary philosophers in relation to especially egregious wrongs (of the sort that defined the 20th century). Philosophers like Bernard Williams have spoken of the "morally impossible," the "unthinkable," etc., to demarcate that which rightfully should stand beyond practical deliberation. This is how personal pacifism understands the refusal to kill.

[2]/ Personal Pacifism's Opposition to War

What of personal pacifism's opposition to war?

Mindful of the dangers of generalizing, one might identify personal pacifism with an individual solution to war over institutional ones. But this need not make opposition a purely personal matter. The emphasis on beginning with one's own personal actions in refusing any involvement with war is in part that by doing so one engages in acts of personal witness that are themselves means to convincing others. One speaks *through* one's actions, as it were. And what one seeks to elicit is a personal transformation in others. This evokes its emphasis on the spiritual, broadly conceived. A common theme is that the changes required to end war must be grounded in a deeper transformation of individuals and communities to be enduring. Thus, Dr. King spoke of the "moral revolution" required.

I think the most significant differences among personal pacifists have involved the question of power. For some, the critique of power implicit in the condemnation of killing has implied a meta-suspicion of the political realm generally. This is how I understand the pacifist religious groups like the Mennonites that emerged from the Radical Reformation. If it is not a meta-renunciation of all political concerns, it has meant the belief that commitment to peace means separating oneself from the larger community or certainly its warlike dimensions. By contrast, the Quakers are commonly noted for how their personal pacifism has meant deep involvement in movements for social justice. The significance of the 20th century's nonviolent movements is in how they have proposed an *alternative* to

---

11 Nagel, Thomas. "War and Massacre." *Philosophy and Public Affairs*, 1, 1972, pp. 123–144.

traditional conceptions of political power that allowed for a principled involvement in the political realm.

[III]/ The Two Types of Pacifism: Political Pacifism

Political pacifism's focus is social institutions. It condemns war as a social practice much as the abolitionists condemned slavery as a social practice, or as many today condemn capital punishment as a social practice. So political pacifism's condemnation of killing is not a condemnation of killing *per se*; rather, it is a condemnation of the *kind* of killing that war as a social practice involves, just as today's opponents of capital punishment condemn the kind of killing that it involves. What political pacifism condemns is the *political* killing (killing by states) distinctive of war, just as what capital punishment opponents condemn is the *legal* killing (killing by government) distinctive of that practice. But just as those who oppose the death penalty may permit killing in other circumstances, political pacifists may permit it as well. Specifically, political pacifists do not regard their opposition to war as implying an opposition to killing in personal self-defense, anymore than opponents of capital punishment regard opposing capital punishment as implying opposition to killing and personal self-defense. In both cases, opponents would say that the institution has nothing to do with personal self-defense.

Like the personal pacifist, the political pacifist regards war as an illicit use of power. But their frameworks are different. Political pacifism's origins are secular and are primarily in 18th-century republicanism. It basically extended the critique of empire in thinkers like Paine to war generally. Its affinity with abolitionism is that abolitionism developed republicanism into a critique of the "slave power" while political pacifism developed it into a critique of the "war power." It is no surprise that republicans like Charles Sumner saw them as variations on the same theme. But in contrast to personal pacifism's concern with "idolatry," the republican concern here is with "corruption." The danger is not political power *per se* but concentrated power, which requires a life of its own by inducing passivity in the community. Hence, the challenge is conceived as one of reclaiming agency over the forces of domination—in this case, war.

Political killing is killing in the service of illicit power, whose full character is revealed by the analysis of the war system.

The first question for political pacifism is *why* it condemns the war system. In contrast to personal pacifism's single-factor condemnation of war (from its condemnation of killing), political pacifism's condemnation is not reducible to any one consideration. Robert Holmes has characterized pacifism as "the principled opposition to war" and war as a "practice that is a detriment to humankind" which he identifies with "the constellation of social, political, economic, religious and ethical practices and values necessary to being able to wage war effectively."[12] The key term here is *constellation*. There are multiple factors involved, as there are with any system of oppression.

I turn to these shortly and to the further question of how political pacifism opposes the war system.

Some further contrasts are worth noting. Personal pacifism approaches matters from the *bottom-up*. It begins with the individual action (killing) and assesses social practices accordingly. Political pacifism proceeds, as it were, from the *top-down*. It begins with the social practice (the war system) and assesses individual actions accordingly. This is a familiar contrast in social theorizing, and the arguments pro and con are familiar. The case for beginning with social practices is that individual actions are always contextually determined, even in their very identity. What constitutes "killing" cannot in fact be determined without presuming a good deal of such context. The case for beginning with individual actions is that social practices are in fact abstractions independent of the actions that constitute them, if only because it is through individual actions that they are reproduced.

Another contrast is that personal pacifism's approach is an *interpersonal* one. Its condemnation of killing is grounded in what it is like to kill another person; it starts on the experiential level, or the encounter between potential killer and potential victim. (Consider the appeal to personal experience by Garrison or Tolstoy.) Political pacifism's approach is, as it were, an *impersonal* one. Its third-person perspective employs the language of rights, which is one reason why it is more congenial to political reflection today. So these represent fundamentally different ethical orientations, where the considerations pro and con are also familiar. The intersubjective approach seems overly subjective, too subject to the vagaries of how experience is interpreted. It seems tangential to the purposes of social policy. The impersonal approach seems overly abstract, especially given how personal—if not intimate—the stakes can be in war. It is this abstraction that so much antiwar literature recounting the personal experiences of combat protests against.

But any consideration of these contrasts must recognize the fact that major figures like Gandhi and Dr. King have somehow merged both orientations.

The reasons for joining them may involve what is necessary for a social movement. Pacifism is not just a theoretical stance and a practical approach. The demands that it makes on activists would seem to require *caring* about peace

12 Holmes, *Op. cit.*, p. 21.

in ways that are not engaged by more structural/impersonal orientations. Such caring involves a willingness to sacrifice and a capacity for hope that the more spiritually laden personal pacifism seems to engender. Personal pacifism is often contrasted with more "pragmatic" approaches. But an overview of the movements for peace suggests that they are often driven by the energy that more personal orientations provide. I speak to some of these issues in the final chapter.

[1]/ The Structure of the War System

It is political pacifism that gives us the notion of the war system. I turn now to a rather schematic account of its elements, and then to how political pacifism critiques and confronts it.

[a]/ War Making and War Building

Political pacifism's starting point is that the key to understanding war and its centrality to modern life is that "war" must be understood not as an event, or a series of events, but as a *practice* possessing a distinct structure and history—that of the *war system*.

The key to understanding the war system is its twofold character that I call *war making* and *war building*.

War *making* consists of the acts of killing and dying in war. War *building* consists of the institutions of mobilizing the human and material resources for the purposes of war making. Analyzing the war system means analyzing both dimensions and their relation to each other.

War making is most identified with *battle*, the "sharp end of war." Carl Von Clausewitz was the first theorist to approach war as a system and the first to explore its inner logic. He is now regarded as an arch proponent of an a-moral "realism," but his unfinished classic *On War* (1832) was a jumble of ideas, some of which were deeply skeptical of war and have inspired critical thinking about war ever since. I draw on some of them in this work.[13]

Clausewitz noted that battle was the most obvious—indeed, the most *spectacular*—dimension of war. For that reason, people are naturally led to equate war *generally* with war making. Note how "war" movies like "Saving Private Ryan" are really "battle" movies. Until recently, studies of war were mainly studies of battles. Sir Michael Howard has characterized this battle fixation as "a kind of historical pornography."[14]

13 Clausewitz, *Op. cit.*
14 See Howard, Michael. "When Are Wars Decisive?" *Survival*, 41(1), 1999, p. 129.

This conception of war-as-battle is central to the ideologies that have always surrounded war.

It informs the entire picture of war as a dramatic confrontation full of the "excitement" that is taken to attract young men to it and compel others to celebrate their deeds. Such dramatic confrontation is the occasion for the display of "martial" virtues valuable in themselves that presumably have no other natural outlet. Victory in war is a matter of which side realizes these martial virtues most fully, meaning that success in war is ultimately a reflection of moral character—of "will," of "sacrifice," and so on. Finally, since the nature of battle is fairly constant over time, war generally is taken to be such. This reification of war leads to a naturalization of it: if war has always been the same, then it seems that war will always be with us.

But this equation of war with battle is a mistake. Indeed, it is the most *fundamental* mistake one can make in approaching the war system.

Clausewitz pressed this as a practical point, pertaining to the conditions of success in war.

The war-as-battle picture conceives war as fundamentally an act of violence, specifically of soldiers killing soldiers. It conceives of war as a large duel, each side trying to kill the other. But while violence is what war most dramatically *involves*, it is not what war is most importantly *about*. Rather, war is fundamentally an act of power, of states compelling states. It is better conceived as a wrestling match, in which each side tries to subdue the other. By conflating war with battle, then, Clausewitz maintained that we fail to grasp what he identified as war's political "essence." All politics is about the exercise of power. War is about violence (its means is killing) in the service of power (its end is compelling).

Put simply, war is *political killing*.

The practical implications of this are several fold. First, it follows that success in war is defined by what happens *beyond* the battlefield—dominating the enemy's "will," in Clausewitz's words. Hence, it is possible to win every battle and lose the war, as the United States learned in Vietnam. And success in war is determined by what happens *before* the battlefield: we *dis*empower an enemy by *over*powering them, by bringing the most human and material resources to the conflict. And this too is a political matter, not of asserting power *against* another society but of exercising power *over* one's own society. Thus, success at war making by states rests on the organization of war building within states. This explains war's impact on society generally, which I turn to shortly.

Thus conceived, war is a decidedly *unromantic* affair. Victory in war has not been principally a matter of heroism on the battlefield. If anything, it has been determined by which states can raise the most revenue to hire soldiers and pay for the weapons. This is evoked in Cicero's famous characterization of money as the "sinews of war." Supremacy in military matters has not been a matter of who has the most courageous soldiers but who has the superior credit rating.

But there is a deeper practical problem with the enterprise itself.

Clausewitz's characterization of war as political has been construed as conceiving war as an instrument for political aims. This is how both contemporary realists and just war thinkers construe it; their disagreement is whether the aims of the instrument should answer to moral concerns as well as strategic ones. But Clausewitz himself does not speak of war as the pursuit of *aims* but rather as war as an extension of *policies*. There is a very great difference between them due to the political nature of the latter. And it bears on the question of whether war is rightly conceived as an "instrument" at all or whether it involves an inexorability of the type I spoke of in Chapter 1.

One political philosopher who has challenged the instrumentalist picture of war is Hannah Arendt. She did so as part of her reflections on the nature of "action." She claimed that war was ill-conceived as an action because it had no *author*. She suggested in *The Human Condition* that monuments to the "Unknown Soldier" bore witness to the "brutal fact" that "the agent of war is actually nobody."[15] By contrast, I take Clausewitz to say that war *has* an author (the polity/state), but it is not a unified, homogeneous subject possessing clear "aims." He suggested that it was composed of a "trinity" of factors, the government, the military, and the people, whose shifting relations with each other made the outcome inherently ambiguous. In the same vein, Michael Mann has characterized the state as a "polymorphous," "promiscuous" entity, or a "patterned mess"—for which talk of "aims" makes no sense at all.[16] War is an extension of policies—but policies are intrinsically inconsistent, unstable, and indeterminate because the polities they reflect are themselves internally conflicted and contested.

That it made no sense to speak of war as an "instrument" of aims was actually a widely held view in the 19th century. It is why many held that it made no sense to speak of wars as just or unjust, but only as whether they were rightly initiated. Political philosopher T.H. Green, in his widely read *Lectures on the Principles of Political Obligation* (1883), held that the corporate nature of the state rendered talk of its "acting" meaningless. When

15 Arendt, Hannah. *The Human Condition*. The University of Chicago Press, 1958, pp. 180–181.
16 Mann, Michael. *The Sources of Social Power: Volume 2, The Rise of Classes and Nation-States, 1760–1914*. Cambridge University Press, 2012, p. 796.

we speak of people killed in war, "(w)e may say that it is by the agency of the state, but what exactly does that mean?"

> No doubt there have been wars for which certain assignable individuals were specially blamable; but even in these cases the cause of the war can scarcely be held to be gathered up within the will of any individual, or the combined will of certain individuals.[17]

At one point, Clausewitz suggested that the main role of "war aims" talk was to create the *illusion* of a unified, homogenous subject where nothing of the sort exists.[18]

Ultimately, then, the practical problem is making war a coherent enterprise at all. Its aimlessness results in an endlessness; its endlessness means that those elements devoted to war achieved dominance within the state, so that the relation of war making to war building is inverted. That is, war building is no longer a means to war making, but war making becomes the means to war building. Note that this danger is appreciated by those policy theorists today concerned with the influence of the "military-industrial complex" on waging war. Their question is "How can states keep war under control?" Pacifists doubt that states can do this for the simple reason that states are themselves ultimately artifacts of war.

[b]/ State Building and Empire Building

If the key to war is its twofold character, and the key to that character has been the primacy of war building, then the key to war building has been the *alliance* of two forms of power: political power, most identified with the state, and economic power, most identified with capital.

[I]/ STATES

These two forms of power may be characterized by their distinctive logics.[19]

---

17 Green, T.H. *Lectures on the Principles of Political Obligation, 1883.* Batoche Books, 1999, p. 121.
18 See Mary Kaldor's discussion of this in "Inconclusive Wars: Is Clausewitz Still Relevant in these Global Times?" *Global Policy*, 1(3), 2010, pp. 271–281; she cites Hew Strachan's "Clausewitz and the Dialectics of War." *Clausewitz and the Twenty First Century.* 1st ed. Edited by Hew Strachan and Andreas Herberg-Rothe, Oxford University Press, 2007, pp. 14–44.
19 I am deeply indebted here to Giovanni Arrighi's *The Long Twentieth Century: Money, Power and the Origins of Our Times.* Verso, 2010.

The political power of the state aims at the domination and expansion of territory by mobilizing and managing the means of violence—or more bluntly, the means of destruction. Its success in this regard is evidenced by its triumph over its competitors as Europe's dominant political form. In 1500, Europe comprised some 500 or so princely domains, independent cities, and contested territories—few of them "states" as we now know them. That they ultimately gave way to a relatively small number of "state"-like forms attests to the superior capacity of the state to wage war against its both internal and external enemies. As Tilly concluded, the European "state structure appeared chiefly as a by-product of rulers' efforts to acquire the means of war."[20]

The economic power of capital aims at the extraction and accumulation of wealth by mobilizing and managing the means of production. Its success in this regard is marked by its triumph over its competitors as modernity's dominant economic form. Thus, state rulers identify power with the extent and populousness of their domains and conceive wealth as a means or a by-product of the pursuit of territorial expansion. Capitalist rulers identify power with the extent of their command over scarce resources and consider territorial acquisitions as a means and a by-product of the accumulation of capital.

This alliance of political power/the state and economic power/capital explains why their stories are so intertwined. Political competition and economic competition have stood in an intimate relation, as manifested in the links between political crises (wars, or crises of destruction) and economic crises (depressions, or crises of production).

For now, let us consider how the two forms of power enable one another.

How did capitalism arise? This is the question of what Marx called "primitive accumulation," and the answer is that it only happened with the violence of the state. Consider a defining feature of capitalist production: the transformation of labor into wage labor or the "commodification" of labor. This did not happen by individuals deciding they were better off working for someone else; rather, individuals were violently dispossessed of their previous means of livelihood via acts like the enclosure movement, and driven to work for others by the alliance of state and capital. This was an internal war that was both occasioned and enabled by external conflicts between states. Theorists of neoliberalism like David Harvey maintain that this state-driven expansion of capitalist relations into other realms is a permanent feature of the capitalist system and not just an artifact of its beginnings.[21]

---

20 Tilly, *Coercion, Capital, and European States: AD 990–1992, Op. cit.*, p. 22.
21 Harvey, David. *The New Imperialism*. Oxford University Press, 2003.

How did the state arise? The parallel to the primitive accumulation of capital was the primitive *consolidation* of the state. Neither of them arose, as their dominant ideologies would have it, as "natural" processes with their own internal logic. Capital was not the natural development of earlier market relations; it triumphed in part by appropriating some and in part by destroying others. Hence, Braudel has stressed that capitalism's relation to the market is one of hostility as much as accommodation.[22] Nor was the state the natural development of earlier forms of rule or government relations—if by "government" we mean the ways in which communities organize their activities. The state triumphed in part by appropriating earlier forms of government and in part by destroying others. Here too, the internal war that this involved was both occasioned and enabled by external conflicts between states.

Thus, in Charles Tilly's famous phrase, "War made the state and the state made war."

We may think of the European state system as originating from a two-fold process. States developed *internally* insofar as their power to dominate became more complete within their own borders. This involved internal struggles of various sorts, involving both negotiation and conflict, resulting in different types of domestic arrangements. States developed *externally* insofar as their power to expand led to an internal imperialism, by which I mean European states conquering their neighbors in a form of quasi-colonization. This continued into the 20th century and is central to its story. Germany regarded Eastern Europe and Russia as fitting subjects of its colonization, to be populated and cultivated by Germans at the expense of "native" populations.

The principal ideology of the state has been that of "protection." Ulrich Beck has written of protection as the "master narrative" of the state, protection against internal enemies ("crime") and external enemies ("war").[23] Feminist scholars have emphasized and explored the gendered dimension of this ideology.[24] Women (and children) are portrayed as the principal ones needing protection, due to their inherently vulnerable status; wars of self-defense are portrayed as protecting wives and mothers (or the "Motherland"). This not only justifies wars but also justifies the social dominance of masculinity as required for war fighting, as the masculine protector puts those defended in a subordinate position of dependence—and obedience.

---

22 Braudel, Fernand. *Civilization and Capitalism, 15th-18th Century*. Vol. III. University of California Press, 1984, pp. 619–633.

23 Ulrich, Beck U. *Risk Society: Towards a New Modernity*. Sage, 1992.

24 See Young, Iris. "The Logic of Masculinist Protection: Reflections on the Current Security State." *Signs*, 29(1), 2003, pp. 1–25.

A racist dimension is added when the threat to women is constructed in racial terms.

At least since Hobbes, this has been conceived as a type of contract. The people alienate their freedom, resources, etc., in return for the protection that the state provides. This "social contract" might be more aptly termed a *war contract*, as its reality is most evidenced in times of war when the stakes are conceived as ones of "survival" and the sacrifices demanded of the populace are the most extreme. Since Rousseau, critics have perceived this contract as a bogus one. Sometimes the threats from which the state presumably protects us are illusory. Mary Kaldor has termed the Cold War as an "imaginary war" in this regard, though a clearer example might be the "War on Terror" in which Western civilization itself was seen as threatened by pockets of terrorism around the globe.[25] But insofar as the threats are real ones, they are generated by the very arrangements—the state system—that claims to protect us from them. Thus, the claim of Tilly, Kaldor, and others that they are best conceived as a kind of protection racket.

[II]/ EMPIRES

Thus far, the picture has been a Eurocentric one. But with the birth of modernity the war system increasingly became a global system, of empires as well as states—or more precisely, of what I shall term state–empires.

Empires long predated the state historically, and modern empires originated independently of the state in the Iberian empires of Spain and Portugal. Their impulses were threefold: strategic, as responses to Islam, proselytizing, as spreading the word of Christianity, and profit—through the extraction of local resources (gold, initially). The latter immediately demonstrated empire's value for war building. Gold extracted from the Americas would fuel Spain's quest for European hegemony.

The immediate upshot was conflict between Spain and Portugal that was addressed in one of the founding documents of the modern world order, the Treaty of Tordessilas (1494). This laid the basis for imperial notions of sovereignty that accompanied the notions of state sovereignty arising within Europe. State sovereignty was inherently limited in two respects: states were bounded on the outside by other states and constrained on the inside by other constituencies, often due to the bargain struck for war making. Imperial sovereignty was inherently expansive: European powers could conquer whole continents in the name of Christianity; indeed,

---

25 Keen, David. *Hidden Functions of the War on Terror*. Pluto Press, 2007; Keen, David. *Useful Enemies: When Waging Wars Is More Important Than Winning Them*. Manas Publications, 2014.

the initial principle held that such powers could conquer entire oceans as well—hence, Portugal claimed the Indian Ocean and Spain the Pacific Ocean. And it was unlimited in the policies of rulers toward the ruled. Initially, those policies were flatly genocidal, and genocide has remained a feature of empire. This provoked a famous debate over the humanity of Indigenous peoples that is regarded as one of the first discussions of human rights. But its upshot was to replace genocide with enslavement, at least as the official policy. The reasoning invoked traditional notions of "just war": Indigenous peoples could not be murdered outright, but, as the vanquished in a "just" war, they could be enslaved. From the start, then, the logic of empire conceived of the relation between the West and the rest in terms of war, be it overt, covert, or temporarily suspended conflict.

This was the ideology of the mercantile empires of the Dutch, English, and French that supplanted Iberian empire and whose conflicts drove the wars of the 17th and 18th centuries. Their novel synthesis of political and economic power was embodied in the "joint stock" companies like the Dutch and British East India companies and England's Royal African Company that were the principal instruments of colonization. They would eventually create imperial structures in which stimulants like sugar and narcotics like opium played a central role. British overseas expansion in the 18th century came to be concentrated around two so-called triangles that effectively covered half the globe. One was in the Atlantic involving slaves, sugar, and manufactured goods; the other was in Asia involving opium, tea, and machine spun cotton. The wealth from empire was especially crucial for British war building as part of its superior ability to finance wars.

The interconnections of these imperial arrangements placed slavery at its heart, to the extent that some argue our entire picture of capitalism needs to be reinvisioned.[26] The capitalism of European states was constructed around wage labor. The capitalism of global empire was constructed around slave labor, which Sven Beckert argues in *Empire of Cotton* began to develop in the 16th century, hence preceded European capitalism and abided by a different logic. War was so central to its logic that Beckert terms it *war capitalism* in contrast to the industrial capitalism that triumphed in Europe. Its core was the violent expropriation of land and labor in Africa and the Americas; it flourished in fields rather than

26 Howard French argues, in his recent book, that the centrality of slavery means that all of modernity must be reinvisioned. French, Howard. *Born in Blackness: Africa, Africans, and the Making of the Modern World, 1471 to the Second World War*. Liveright, 2021.

factories, and it was not mechanized but land- and labor-intensive.[27] The former slave Gustavus Vassa, now commonly known by his African name, Olaudah Equiano, famously defined slavery itself as a perpetual "state of war." Vincent Brown has suggested that another name for imperial arrangements of the 17th and 18th centuries might be the "Atlantic War of Slavery." Bodies were extracted from Africa through outright war, decimating communities there; they were then subjected to warlike measures in the violent institutions of slavery; finally, there were periodic outright rebellions by slaves, the specter of which hung over the entire enterprise.[28]

"It was slavery that made empire pay and empire that made slavery possible," write Jane Burbank and Frederick Cooper.[29] Economically, empire was now seen as a crucial if not essential condition for the growth of capitalism in Europe. It was also politically essential in the strengthening of European states and especially in their capacity to wage war. As noted, a crucial institution was the "joint stock company," which flourished into the 19th century blurring the distinction between economic and political institutions. The first such companies in England emerged from the practice of "privateers," private individuals contracted by the state to engage in war and plunder. The Royal African Company, granted a monopoly in the early 1600s to supply slaves to the English colonies, grew England's share of the slave trade from 33% in the 1670s to 75% in the 1680s. The removal of this monopoly led English slave trading to further thrive so that it carried one-third of all slaves shipped across the Atlantic, dominating the global slave trade prior to its abolition by Parliament in the 1800s. The wealth generated for the English state was crucial to its financial solvency and waging wars.

The formal abolition of slavery in the 19th century paralleled the earlier condemnation of genocide insofar as it did nothing to undermine the principle of imperial sovereignty. Peoples could no longer be directly enslaved, but the colonized could still be compelled to labor in conditions that denied them all civil rights. This was especially true in Africa, which was divided among European powers by the Berlin Conference at the end of the century. Empire's contribution to war building became even more direct, as colonies were no longer principally sources of revenue but sources of soldiers as well. By the early 1800s, the combined strength of British Indian armies was over 150,000, making them one of the largest standing

27 Beckert, Sven. *Empire of Cotton: A Global History*. Alfred Knopf, 2014.
28 Brown, Vincent. *Tacky's Revolt: The Story of an Atlantic Slave War*. Harvard University Press, 2022.
29 Burbank, Jane and Cooper, Frederick. *Empires in World History: Power and the Politics of Difference*. Princeton University Press, 2011, p. 178.

armies in the world. This culminated in their role in World War I, as noted in Chapter 1.

If the principal ideology of the state has been protection, that of empires has always verged on *paranoia*. This has been packaged in different forms. Michael Mann speaks of "securing survival." "The motive of strategic security," he writes, "is usually seen by imperialists as defensive expansion against threats from other states or empires. The bigger the empire, the less secure it feels!"[30] This echoes a remark of Joseph Schumpeter's about Rome: "Rome was always being attacked by evil-minded neighbors, always fighting for a breathing space. The whole world was pervaded by a host of enemies, and it was manifestly Rome's duty to guard against their indubitably aggressive designs." Hence, there was "no corner of the known world" where Rome was safe from attack—and its fighting "was always invested with an aura of legality."[31] The early modern empires framed this in the language of Christian versus Pagan. The specter of Islam was never far from its mind. But in the 19th century, ironically related to the abolition of formal slavery, empire's ideology increasingly became one of racism. Now it was their status as "lesser breeds," in Rudyard Kipling's words, that explained European dominance.

[2]/ The Critique of the War System

The complex nature of the war system means that the critique is a complex one, proceeding on multiple dimensions.

[a]/ The Critique of War Making and War Building

The first way to frame the critique is via the distinction between war making and war building.

It is in the critique of war making that pacifism most overlaps with just war thinking. Over the last century, as civilians have become overwhelmingly the victims of killing in war, attention has especially focused on questions of *jus in bello* and how war is conducted. Do the methods of war making necessarily involve massive violation of individual and communal rights? This was already a concern in the mid-19th century, prompting a series of efforts by states to constrain the crimes of war making. Scholars

---

30 Mann, Michael. *Sources of Social Power: Vol 3, Global Empires and Revolution, 1890–1945.* Cambridge University Press, 2012, p. 21—referring to James. Harold. *The Roman Predicament: How the Rules of International Order Create the Politics of Empire.* Princeton University Press, 2006.
31 Schumpeter, Joseph. *Imperialism and Social Classes.* Ludwig von Mises Institute, 2007, p. 51.

of these international arrangements have portrayed them as responses to pacifist-type doubts about war generally. These concerns came to a head in World War I when late 19th century/early 20th century measures to constrain war so obviously failed to limit the carnage. The turn to international arrangements of war prevention/conflict resolution following World War I reflected the conclusion by some that war could not be morally constrained; hence, the only option was to minimize its occurrences. Pacifist concludes that war making cannot be morally constrained in any significant sense nor is there any reason to think that arrangements allowing for "just" wars have proved viable.

Today's contingent pacifisms respond to these issues of war making. They approach the question from the perspective of individuals contemplating fighting in a war. Their argument is one of moral prudence: given the overwhelming likelihood of violating or being implicated in the violation of *jus in bello* principles, one is morally obliged to refuse involvement in it; similarly, given the overwhelming uncertainty that a war will abide by the principles of *jus ad bellum*, one is morally obliged to refuse involvement in it. I shall raise some questions later about how "contingent" our judgment should be in these matters, but such contingent pacifism is clearly a central pillar of any pacifist approach.

My remarks on the priority of war building over war making suggest that the deeper problems may be found there.[32]

Some of these involve the actions of states *in* war. Consider some of the issues in states mobilizing individuals *for* war:

- Theorists of individual rights have always questioned the fate of such rights in the state's enlistment of soldiers, either by direct compulsion or by exploiting economic vulnerabilities. Recently, attention has been devoted to how the state transforms individuals into soldiers once enlisted.[33] To what extent does it involve practices of psychological coercion and abuse, verging on torture? Note that the traditional argument in defense of such practices is that states could not fight wars without the license to engage in them. Pacifism agrees with this but concludes that this is why war should not be fought.

---

32 For an excellent discussion of the costs of war building, see Dobos, Ned. *Ethics, Security, and The War-Machine: The True Cost of the Military*. Oxford University Press, 2020.

33 See Reese, David "Regulation of Bodies as Gendered Nationalistic Ideology: Physically Wounded Veterans as Political Props." Master's Thesis, University of Oregon, 2015; Robillard, Michael and Strawser, Bradley. *Outsourcing Duty: The Moral Exploitation of the American Soldier*. Oxford University Press, 2022.

- Mobilizing for war has provoked massive civil conflict. France's mass mobilization during its revolution, the *levee en masse*, provoked revolts throughout the country that were generally repressed. But in one region of coastal France, the Vendee, the revolt turned into outright rebellion. It evoked the 14th-century *Jacquerie* rebellion, when French peasants revolted against the conditions of the One Hundred Years War, though now the revolt became explicitly counterrevolutionary and royalist. Its cost was a bloody one: the War in the Vendee (1793–1796), left over 200,000 dead, approximately 20–25% of the territory's population. Debate surrounds whether it is properly termed a genocide. The New York City Draft Riots (1863) remains one of, if not the most serious civil disturbances in American history. The racially charged conflict was provoked by the draft of men for the Civil War; President Lincoln had to divert troops from the recent Battle of Gettysburg to quell the rebellion.
- War mobilization has invariably meant the infringement on if not abolition of civil liberties when they existed. In the United States, the very *notion* of "civil liberties" as a distinct concern arose in response to the government's repressive policies during World War I.[34] The aftereffects of that war directly led to the racial terrorism that characterized the 1920s. This racialized atmosphere led directly to the internment of Japanese Americans in World War II, one of the most egregious civil liberties violations in American history. And the actions of the American government were mild compared to others.
- Finally, there are the genocides resulting from the "security" needs of war. The Armenian genocide (1915–1917) involved the killing of 800,000–1.5 million Armenian men, women, and children. Its causes were complex and still debated. But central to them was the stigmatization of Armenians as a hostile population amidst a country mobilized for war; the genocide was in fact enabled by the number of Armenian men conscripted into the military hence absent for resistance. German Nazis would adopt it as a model for their own genocidal policies.[35]

All of this pertains to what states *do*, in the process of war building. But the deepest problems involve what states fundamentally *are*—as agents of war building.

If one had to generalize across the millennia, the best characterization of intergroup violence would probably be that it has been basically a form of *theft*. The great liberal thinker Benjamin Constant (1767–1830) gestured

---

34 Murphy, Paul L. *World War I and the Origin of Civil Liberties in the United States.* Norton, 1979.

35 Shaw, Martin. *War and Genocide: Organised Killing in Modern Society.* Polity, 2003.

toward this in his classic essay, "The Liberty of the Ancients and the Liberty of the Moderns" (1816). Speaking to whether there was any feature that distinguished war across the ages, he wrote:

> War and commerce are only two different means of achieving the same end, that of getting what one wants. Commerce is simply a tribute paid to the strength of the possessor by the aspirant to possession. It is an attempt to conquer, by mutual agreement, what one can no longer hope to obtain through violence. A man who was always the stronger would never conceive the idea of commerce. It is experience, by proving to him that war, that is the use of his strength against the strength of others, exposes him to a variety of obstacles and defeats, that leads him to resort to commerce, that is to a milder and surer means of engaging the interest of others to agree to what suits his own.[36]

What Constant is describing is violence as a form of blatant *extortion*. The critique of the war system as essentially a form of protection racket takes it to be a form of *covert* extortion. As I unpack this critique, there are parallels with Marx's critique of capitalism. Marx held that the illusion of a contract between capital and labor obscured how capitalism was fundamentally a system of exploitation. The claim that "war is a racket" holds that the illusion of a contract in which states receive human and material resources for "protection" obscures how the system is fundamentally one of extortion. In both cases, it should be stressed, the claim is that this is a fact about the structure of the system and not about the designs or motives of those in charge of it. There is an *impersonality* to it, which raises a further aspect of the critique.

[b]/ Injustice and Inhumanity

To speak of war's "injustices" minimizes the problem, as I understand it. It frames matters in terms of one individual's/group's violating another's rights. By *inhumanity*, I mean to evoke the problem raised by the system's inexorability, its penchant to escape all human control including that of those who violate the rights of others. I have noted that this problem became explicit with the experience of World War I (one manifestation of this was the theme of "totality" in philosophy, as designating an all-inclusive and self-propelling system that resists all standard critiques). It speaks to

---

36 Constant, Benjamin. *Constant: Political Writings*. Translated by Biancamaria Fontana, Cambridge University Press, 1988, p. 313.

the fate that all share who are implicated in the war system, hence one that can only be escaped by abolishing the system.

Critics of this dimension have noted the element of absurdity here. There is, after all, something absurd about a system that manufactures the very insecurities it claims to resolve; it is, as someone once said of Freudianism, the disease for which it purports to be the cure. Notions of "madness" invariably suggest themselves. Henry David Thoreau suggested that states belonged in an "asylum" along with all the other persons who conjure up imaginary threats and fantastic schemes to solve them. Plus, there is the irony of fate in the dimension of what I have termed the "chickens coming home to roost"—how crimes that some inflict on others are ultimately inflicted on them. A case in point is how the fate of Europe in World War II has been diagnosed by anti-colonialist voices as basically Europeans receiving their own medicine. Franz Fanon wrote in *Wretched of the Earth*, "What is fascism if not colonialism when rooted in a traditionally colonialist country?"[37] In his *Discourse on Colonialism*, Aime Cesaire spoke of how the crimes in Europe reflected "a terrific boomerang effect" in which violent policies that "until then had been applied only to non-European peoples" now made victims of past accomplices. Cesaire angrily announced that what offended Europeans was not the crimes *per se*, but the fact that crimes are committed which until then had been reserved exclusively for the Arabs of Algeria, the "coolies" of India, and the "niggers" of Mrica.[38] So, everyone will escape such fates with the abolition of the war system.

The alienation here—of individuals and groups from their own agency, of their subjecting themselves to an impersonal fate—ultimately refers to the logic of *power*. This critique of power marks a convergence of personal and political pacifism, though with different diagnoses. Personal pacifism's religious background frames the critique of power in the language of *idolatry*. War is a matter of false worship by which persons, in claiming godlike powers over others, in fact renounce capacities that render them fully human. Political pacifism frames it in the language of *corruption*, which is most directly a legacy of republicanism. Power plays on the inherent weaknesses of human character that can be addressed by civic virtue but ultimately requires rightly ordered institutions to keep power in check. I consider later how this bears on the task of nonviolence.

---

37 Fanon, Frantz. *The Wretched of the Earth*. Grove Weidenfeld, 1963, p. 89.
38 Césaire, Aimé. *Discourse on Colonialism*. Monthly Review Press, 2001, pp. 36–37.

[3]/ Confronting the War System

I now turn to the sort of practical measures pacifism has endorsed to oppose the war system. Let me begin with a general point.

Given how all-encompassing the war system is on this account, given how baked into our institutions war is, the question of how to challenge it has no simple, single, or fixed answer. The task of challenging it can seem overwhelming; it is like addressing climate change in this regard. As with climate change, there are almost too many things to do. Not surprisingly, political pacifists have disagreed about this matter. One thing that unites them is a commitment to an ongoing dialogue about strategies of opposition which will include the fact that the war system changes in ways that opposition to it must address. Resistance can be spontaneous, but it takes a long time to develop political strategies. Organized efforts against war are not even two centuries old, and the systematic study of them is barely beginning.

[a]/ Peace Making and Peace Building

Again, the first way to frame a discussion of opposing the war system is via the distinction between war making and war building. Accordingly, we may identify political pacifism's practical approach as involving two dimensions: *peace making and peace building.*

Peace making is what most people think of when they think of opposing war. It mainly pertains to opposing *particular* wars, though it can involve doing so by institutional means. We may usefully distinguish three aspects of it.

- *Preventing wars before they occur:*

A feature of the modern war system starting in the 19th century is people's awareness of the growing momentum for war. Sometimes this means awareness of the machinations of leaders, as when the Bush administration signaled its intention to invade Iraq long before it actually happened. Peace making involved the massive global demonstrations against the impending action. Sometimes it means awareness of contending forces at work that signal the imminence of conflict. One of the most dramatic examples of peace making was the feverish attempts by pacifist leaders like the French Jean Jaurez to head off World War I. Further institutional expressions of this have been attempts to induce more caution in states before they go to war. Examples of this include the treaties promoted by Secretary of State William Jennings Bryan before World War I in which states agreed to submit their conflicts to mediation at least for a certain period of time or to cooling off period prior to armed conflict.

• *Opposing wars when they occur:*

This involves ongoing protest against and other forms of resistance by the citizenry to wars once they have started. World War I saw the first dramatic examples of this. The development of mass armies has meant that opposition to ongoing wars has come to include soldiers themselves. The actions of American soldiers during the Vietnam War are among the most dramatic examples of this. But there are more subtle forms of resistance in the forms of passive refusal. One thinks of the much-celebrated "Christmas truce" of December 1914 where, during an official cease-fire, French, German, and British soldiers on the Western Front left their trenches to exchange season greetings, swap food and souvenirs, play football, and join together in caroling. Such truces were part of a larger "live and let live" system of informal agreements among soldiers to refrain from shooting at each other at specific times, such as when they exercised or worked in view of the enemy. Military leaders saw this as undermining their efforts and worked to oppose it.[39]

Such spontaneous resistance has loomed large in the cultural memory of antiwar activism. William Faulkner's Pulitzer Prize winning novel *A Fable (1954)*, inspired by the widespread mutiny of French soldiers on World War I's Western Front, recounted their heroism and then the martyrdom of a Christ-like figure who stood up against the insanity of all war. It helped inspire the subsequent film, Stanley Kubrick's "Paths of Glory" (1957), a similar story of martyrdom based on the French army's "Souain corporals affair" when four innocent men were court-martialed and shot as an example to others.

• *Ending wars after they have begun:*

The most striking examples of this involve international efforts by citizen groups. After the onset of World War I, but prior to America's entrance into it, women peace activists organized an international congress of women involving that met at The Hague in April 1915. In one of the remarkable episodes of peace making, they committed themselves to personally engaging the leaders of the belligerent countries to encourage them to seek a mediated piece. Its leaders including Jane Addams ended up meeting with, among others, British Prime Minister Asquith and Foreign Minister Grey in London, and German Chancellor Bethmann Hollwegg and Foreign Minister Jagow in Berlin. When they met with Austrian Prime

---

39 Ashworth, A.E. *Trench Warfare 1914–1918: The Live and Let Live System.* Palgrave Macmillan, 1980.

Minister Karl von Stuergkh, he responded to the concern that their mission might seem foolish by saying that theirs were the first sensible words he had heard since taking office.

These are all instances of peace *making*. One may think of peace *building* as addressing war at the more institutional level. If war is truly baked into our most fundamental institutions, then war's momentum will only be checked by challenging its institutions—by creating an alternative *peace system*. Yet fundamental change does not occur in one fell swoop; on the contrary, it seems invariably incremental.

Again, we may distinguish three aspects.

- One aspect of peace building is constructing alternative arrangements to resolve the problems posed by war without resorting to war.

Two examples are arrangements that may involve violence but do not in the eyes of their proponents involve war. I have anticipated this in my remarks on disagreements about what constitutes "war." A first example is popular arrangements of self-defense that do not involve organized armies but do involve the capacity for violence. This is what Jean Jaurez proposed as replacing armies and war, as he conceived them. A second example is arrangements of collective security related to something like a global government. Here too, the claim is that such international law enforcement does not constitute "war." Proponents of both feel that this is not a purely semantic issue but has to do with the realities of the war system, as detailed here.

- Another aspect of peace building is reconstituting our institutions so that they do not invariably carry us into the conflicts of war.

One can place this concern at the center of the religious movements for pacifism that in Christianity since the Reformation have had a marginal but strongly enduring presence. Theirs has been mainly an approach of constructing alternative communities seeking a spiritual core at their center that does not generate the kind of animosities that secular arrangements invariably do. But these have always had an impulse to separate themselves from larger society that they regard as, at least for the time being, hopelessly corrupt.

The problem this poses for the war system will become clearer from the next chapter's consideration of how states, empires, and nations have been constituted by war. William James, in his classic essay "The Moral Equivalent of War," approached it in terms of how much our existing political institutions required war for their *vitality*. Another issue is how much the inherent logic of those institutions is such as to promote war, which they then appeal to as their justification—that is, the "protection racket" logic. Thus, some have challenged the state from an anarchist perspective, proposing more mutualist-type arrangements as ones that neither presume nor promote endless conflict. Others have challenged the nation by proposing to replace it with notions like "country," whose substance they regard as less artificial and hence less given to affirming their reality in the enterprise of war.

- Finally, I would highlight how peace building as reconstituting our institutions must invariably involve re-envisioning our *culture*.

I take this to be the animating thought that peace building is ultimately a matter of peace education. But "education" in the broadest sense: as a collective, ongoing dialogue—one that is attentive to the ways in which violence constitutes our most fundamental relations to each other and conceptions of ourselves; and is inquisitive of all the ways these might be reconstituted in a vision of peace. For example, Stanley Hauerwas has argued that replacing the war system means nothing less than a re-conception of history and our relation to it.

War possesses our imaginations, our everyday habits, and our scholarly assumptions. When we begin to think what it might mean to abolish war, we cannot help but call into question some of our most cherished convictions—convictions whose importance we may not recognize because they so seldom need to be made articulate. Enacting and narrating present and future peace will entail many spiritual, moral, and even aesthetic demands.[40]

[b]/ Reconceiving Power

Another way to frame the discussion of opposing the war system is via the distinction between injustice and inhumanity.

How do we fashion practices of resistance to the war system that do not contribute to its inhumanity, construed as its penchant to escape human control? This is the question of what it means for such practices to fully succeed, not just now but in the long run. For example, alternative security arrangements must not just resist aggression but do so in a manner that they themselves take on a life of their own. This concern acts as a constraint on our approaches to peace making and peace building.

---

40 Hauerwas, Stanley, et al. "The Case for Abolition of War in the Twenty-First Century." *Journal of the Society of Christian Ethics*, 25(2), 2005, pp. 17–35.

This ultimately raises the question of *power*, specifically what it means to *demilitarize* power—as the project of nonviolence aspires to do. Anticipating the historical discussion of the next chapter, I would place the problem this way. Feudalism's demise and the emergence of modernity reflected a dramatic expansion and concentration of *violent* power, linked to dramatic changes in military technology. This both enabled and was enabled by a new institution, the state, to manage such powers; subsequent changes in the state reflected changes in the techniques of violent power. The conclusion would seem to be that unraveling the war system involves a more *radical* change in the nature of power itself, as proponents of nonviolence have maintained. But what this means requires a deeper exploration of nonviolent power, both theoretically and practically. It means conceiving of nonviolence not just as another tactic of power but as engendering its fundamental transformation.

Theorists like Jonathan Schell have found value in Hannah Arendt's discussions of these matters, specifically her view that violence and power abide by distinct if not contrary logic. Violence, as the physical infliction of force, is instrumental in character, to be assessed by whether its means would achieve its ends. As such, there is nothing even particularly human about it. By contrast, power as Arendt understands it "corresponds to the human ability not just to act but to act in concert. Power is never the property of an individual; it belongs to a group and remains in existence only so long as the group keeps together."[41] Its choral nature means that power is to be assessed by its legitimacy, determined by how people came to act in concert. The conclusion that theorists like Schell find compelling is that, whereas violence may defeat power, legitimate power can only be nonviolent; only such legitimate power will ultimately endure.

In my view, this defines an agenda that a full analysis of the war system must address. How did the distinction between violence and power in fact become so blurred, not just theoretically but practically, that the nonviolent dimensions of power became all but invisible? What are the subtle ways in which nonviolent power can still be implicated in violence? And how do we reconceive notions of legitimacy in relation to power that is truly nonviolent—for example, how does its legitimacy involve giving full expression in human agency rather than rendering such agency a tool for clean purposes?

[IV]/ The Problem of Absolutism

Let me return now to the problem noted earlier about the apparent "absolutism" of pacifism's unconditional condemnation of war.

---

41 Arendt, Hannah. *On Violence*. Houghton Mifflin Harcourt, 1970, p. 44.

The specific problem I shall address is what it means to absolutely condemn a *system*, as in political pacifism's condemnation of the war system. Does the condemnation of the war system imply condemning every *instance* of war, past and future? Since war generally has been condemned from many quarters in the past century, I have suggested that what distinguishes pacifism is its unqualified condemnation, meaning that it does not make exceptions for past and future cases.

So, some defense of this unqualified condemnation is required against the charge of dogmatism.

But before considering the worries about it, some words are in order about the *merits* of such positions, if only so that we don't throw out the baby with the bathwater. Absolutist positions have often been the agents of needed social change. This was true of 19th-century abolitionists, whose members were attacked as dogmatic fanatics. I am old enough to remember when the 1960s Civil Rights Movement was critiqued on the same grounds for slogans like "Freedom Now!". The worry about absolutism is its apparent illiberalism. Yet liberals themselves have noted its value. Walter Lippmann cited William Jennings Bryan in this regard. "William Jennings Bryan once said that to be clad in the armor of righteousness will make the humblest citizen of all the land stronger than all the hosts of error."[42] It was on such grounds that Jane Addams voiced a pragmatic-type argument for absolutism in her *Peace and Bread in Time of War*:

> My temperament and habit had always kept me rather in the middle of the road; in politics as well as in social reform I had been for "the best possible." But now I was pushed far toward the left on the subject of the war and I became gradually convinced that in order to make the position of the pacifist clear it was perhaps necessary that at least a small number of us should be forced into an unequivocal position.[43]

The worry about absolutism's apparent illiberalism arises from belief in the importance of remaining open minded, flexible, etc., especially in matters of deep human importance—like war. This concern needs honoring as the opposite orientation so apparently leads to the commission of violence, if not war. But let me suggest how the experiences of the last century have complicated the issue. What I have in mind is the fact that one response to the horrors of the last century from many quarters has been that they reflect a *decline* of absolutist prohibitions of the sort that should render

42 Lippman, Walter. *The Public Philosophy*. Routledge, 2017, pp. 137–138.
43 Addams, Jane. *Peace and Bread in Time of War*. University of Illinois Press, 2002, p. 77.

certain actions not just morally impermissible but morally impossible. We find this in such disparate thinkers as Hannah Arendt and Albert Camus.

We also find it in thinking about the laws of war in contemporary just war theory, particularly in its approach to principles of *jus in bello*. The deontological liberalism that rose to prominence in the last part of the 20th century, inspired for some by alarm at America's crimes in the Vietnam War, insisted that there were absolutist-type limits on what could be done in war. This issue was starkly raised again some years later by the practices of torture employed in America's "War on Terror."[44] So, literature proliferated aiming to *reclaim* such prohibitions. This was not, it should be stressed, incidental to the general question of war. On the contrary, I think it came from the recognition that *only* if such absolutist constraints could be established could the legitimacy of war generally be confirmed. The larger politics of this has been explored by Samuel Moyn in how a case for "humanitarian war" has been central to legitimizing the kind of wars that America has fought in recent decades.[45]

It would seem, then, that the difference between pacifism and such just war approaches is not absolutism *per se*, but what they are absolutist *about*. This does not resolve the problem of absolutism, but it does suggest that a conundrum for just war thinking is that, in its rejection of pacifism, it cannot employ arguments against pacifist absolutism that do not at the same time undermine its own absolutism. For example, a standard objection to pacifism appeals to a "lesser evil" argument. It argues that refusing to condone a particular war might lead to much worse consequences overall. Again, this is a serious consideration. But such "lesser evil" arguments can also be raised against the constraints that just war thinkers endorse in war, that is, that abiding by them might lead to much worse consequences overall. So if the just war theorist wants to fault pacifism for its absolutism, the challenge they face is how to do so without compromising their *own* absolutism—which they regard as necessary for the legitimacy of war itself.

I have so far posed the problem as one of "absolutism." But I think this is actually a bad term. It mainly suggests that pacifism/absolutist positions presume claims to "absolute truth." But this is certainly not true of pacifism as I am construing it; on the contrary, it insists that understanding war— hence understanding what it is that one rejects—is an empirical matter,

44 Mayerfeld, Jamie. "In Defense of the Absolute Prohibition of Torture." *Public Affairs Quarterly*, 22(2), 2008, pp. 109–128; Waldron, Jeremy. "Torture and Positive Law: Jurisprudence for the White House." *Colombia Law Review*, 105(6), 2005, pp. 1681–1750.
45 Moyn, Samuel. *Humane: How the United States Abandoned Peace and Reinvented War.* Farrar, Straus and Giroux, 2021.

whose judgments are grounded in attention to history and institutions hence could be revised in that light. The terms "absolutism"/"absolutist" strike me as loaded from the start. So instead, let us characterize pacifism as an *uncompromising* position. It is *theoretically* uncompromising in its condemnation of war, and it is *practically* uncompromising in its opposition to war.

Framed thus, I shall take pacifism's uncompromising condemnation of war as raising two concerns.

One is that it is too *categorical* with regard to past instances of war. Its blanket judgment ignores the particularities of those past conflicts that render some of them exceptions to its negative appraisal. The other is that it is too *presumptuous* with regard to future instances of war. Its blanket judgment ignores the particularities of new conflicts that might render some of them exceptions to its negative appraisal. These worries are related in that if there are exceptions to pacifism's blanket judgments about past wars, then there may be exceptions in future wars.

It is to accommodate these worries that some have endorsed more compromising forms of pacifism. For example, they would relax pacifism's unconditional condemnation of past cases to allow that some good wars may have happened in the past; typically, the exception they allow for is World War II. This was the position of Albert Einstein and Bertrand Russell, who otherwise identified themselves as pacifists.

We might say that their condemnation of war is a *general* one but not a *universal* one insofar as it allows for contrary cases. But since contrary cases *could* happen again, open-mindedness requires that pacifism assess each new war as it comes along.

This is one form of what is now called "contingent" pacifism, the exploration of which is a significant development of recent pacifist thinking.[46] It is a "contingent" pacifism in the sense that its judgment of war is contingent on further facts about particular wars as they present themselves. But since "contingent pacifism" has several meanings, I shall prefer to term it *provisional* pacifism. I shall remark on the differences between pacifism as I construe it and such provisional pacifism, but let me remark on what rests on this. I have noted the demarcation problem between pacifism and a fastidious just war approach. A similar question is how much actually rests on the distinction between uncompromising

---

46 May, Larry. *Contingent Pacifism: Revisiting Just War Theory*. Cambridge University Press, 2015; Parkin, Nicholas. "Conditional and Contingent Pacifism: The Main Battlegrounds." *Critical Studies on Security*, vol. 6, no. 2, pp. 193–206; Bazargan, Saba. "Varieties of Contingent Pacifism." *How We Fight*. Edited by H. Frowe and G. Lang, Oxford University Press, 2015, pp. 1–17.

and more compromising forms of pacifism, specifically what is the *practical* difference? Provisional pacifism, like a fastidious just war approach, may imply such a massive presumption against war that it is hard to see the practical difference between this and total condemnation. It is like the difference between someone who rejects all capital punishment and someone who imposes standards of proof on it that in practice may never be met. I think the main difference is what each implies about *preparations* for war: does granting the "possibility" of just wars commit one to preparing for them (in ways that may in fact promote *unjust* wars)? I shall say more about this issue as we proceed.

[1]/ Too Categorical?

Is uncompromising pacifism too categorical in its condemnation of past wars? I think that the general disillusionment with war today means that this worry takes a quite *specific* form. Specifically, it raises the World War II problem: when pacifists say they categorically condemn war, they are challenged with "What about World War II?" Note that there have been other conflicts in the 20th/21st centuries that many believers in just wars have regarded as just. Many regard the United Kingdom's actions in the Falkland War as just, or the First Gulf War as just. But the response to the pacifist is typically *not*: "What about the Falkland War!" or "What about the first Gulf War!" But: "What about *World War II*?"

I think this is because, whatever their justice, the Falkland War or the first Gulf War are not regarded as significant enough to rest a *general* case for war on—given the widespread skepticism of the endeavor. Yes, there were reasons to fight those particular wars, but doing so served to sustain a practice that is generally immoral; indeed, it is *so* immoral that the moral *dis*value of sustaining the practice could be seen as outweighing the moral value of fighting in these exceptional cases. So let Argentina have the Falklands, let Iraq have Kuwait: it is not worth sustaining what is widely regarded as an immoral practice.

But World War II leaps out because it does seem so different in that the stakes were so high. I would put it this way: it is not just an exception to the unconditional condemnation of war; it is a *super-exception* to it—of the sort that renders pacifism's ignoring it dogmatic. A "super-exception" to a generally immoral practice is an act whose performance is of such great moral urgency that it outweighs the moral disvalue of sustaining the practice generally. All practices that we deem immoral have exceptions; the problem arises with super-exceptions. Arguments against uncompromising positions typically invoke *hypothetical* super-exceptions: "Would you torture someone if the fate of the earth depended on it?" World War II is proposed as an actual super-exception.

My first response has been anticipated in Chapter 1. What do we *mean* by "World War II"? The appeal to World War II is highly selective in what is included and what is not. The good parts are picked out and the bad parts ignored, an approach made easier by the very diffuseness of the conflict due to its imperialist character.

But *anything* can be justified if we can ignore the bad parts and focus only on the good ones.

I have said that when people speak of World War II as the "good war," what they generally have in mind is the conflict in Western Europe between the Allies and Germany. They ignore other aspects of the war like the Bengali famine. Nor do they have in mind much that happened in the Pacific War. In 1938, to stop Japanese advances on Wuhan, America's ally Chiang Kai-shek breached the dams of the Yellow River to flood the surrounding area in what is the greatest act of environmental warfare in history. After first blaming the Japanese, Chiang Kai-shek acknowledged that 12 million people had been impacted by the flood but portrayed it as evidence of the Chinese "will to resist." Current scholarship places the deaths due to the flood as between 500,000 and 900,000, and the total refugees between three million and five million. It continued to cause economic and environmental disarray for years after the war.[47] (Most experts regard Chiang Kai-shek's actions as having achieved nothing militarily, as the Japanese seized Wuhan four months later by attacking from a different direction.)

Recent scholarship has stressed this point in talk of the "good war." (A profound meditation on the war from a pacifist perspective is Nicholson Baker's *Human Smoke: The Beginnings of World War II, the End of Civilization*.[48]) What is striking about World War II is how much is taken to rest on its example versus how much is really known about it. Is our ignorance of it a reflection of the ideological weight it has come to bear? This is a point raised by Paul Fussel, himself a decorated World War II combat veteran. He writes, "For the past fifty years the Allied war has been sanitized and romanticized almost beyond recognition by the sentimental, the loony patriotic, the ignorant, and the bloodthirsty." He adds,

Now, fifty years later, there has been so much talk about "The Good War," the Justified War, the Necessary War, and the like, that the young

---

47 Dutch, Steven. "The Largest Act of Environmental Warfare in History." *Environmental & Engineering Geoscience*, 15(4), 2009, pp. 287–297; Lary, Diana. "Drowned Earth: The Strategic Breaching of the Yellow River Dyke, 1938." *War in History*, 8(2), 2001, pp. 191–207.

48 Baker, Nicholson. *Human Smoke: The Beginnings of World War II, the End of Civilization*. Simon and Schuster, 2008.

and the innocent could get the impression that it was really not such a bad thing after all. It's thus necessary to observe that it was a war and nothing else, and thus stupid and sadistic, a war, as Cyril Connolly said, "of which we are all ashamed" . . . . further, a war opposed to "every reasonable conception of what life is for, every ambition of the mind or delight of the senses."[49]

But the biggest problem is separating World War II from World War I. Does it make any more sense than if someone responded to critiques of the Thirty Years War by saying: "But let's just focus on the Palitanate Campaign (1620–1623)—surely the Protestant Union was justified in resisting the Spanish Empire!" If World War II is detached from World War I, then the conclusion is that war can involve ultimate stakes. If the two are appraised together—as they must be, in my view—then the absurdity of war generally includes the fact that it generates conflicts containing the delusion of "ultimate stakes."

Prior to World War II, some argued against America's involvement by appealing to its larger implications for the war system. These voices have been relegated to the dustbin of history but their arguments cannot simply be dismissed. The great historian Charles Beard held that the long-term consequences of America's involvement would be sustaining if not expanding a militarism that outweighed any good that the war would achieve. In retrospect, World War II led to the creation of a massive permanent military establishment whose signal achievement was the development and proliferation of nuclear weapons threatening the fate of the earth. Since then, every war America fights or endorses has been packaged as a super-exception whose occurrence necessitates continuing preparations for future super-exceptions. Korea was another World War II, Vietnam was another World War II, the conflicts in Central America were more World War II's, the "War on Terror" yet another one, as is the current "Great Power Conflict."

[2]/ Too Presumptuous?

Is uncompromising pacifism to presumptuous in its condemnation of future wars? Does open-mindedness oblige us to consider each new case as it comes along? Let me raise some further issues related to this concern.

---

49 Fussell, Paul. *Wartime: Understanding and Behavior in the Second World War*. Oxford University Press, 1990, p. 142; Samet, Elizabeth D. *Looking for the Good War: American Amnesia and the Violent Pursuit of Happiness*. Farrar, Straus and Giroux, 2021; Sherry, Michael. *The Rise of American Air Power: The Creation of Armageddon*. Yale University Press, 1989.

[a]/ Is It Rational?

A first question is whether open-mindedness to each new case that comes along is really *rational*.

Scholars identify 14,600 wars in recorded history and 3,200 major ones. Yet, as just war theorist Jeff McMahan has noted, even the most knowledgeable people about war disagree among themselves about the justice of virtually every one of those wars.[50] For the sake of argument, let's assume that there is general agreement about the justice of *one* of those wars— World War II. So that is 1 out of 14,600, or about 0.00007% of all wars.

Given that record, is it really rational to spend the time assessing each new war that comes along, given the minuscule chances of there being another "just" one? Based on experience, the chances of another just war coming along are significantly less than the chances of the earth being destroyed by a meteor.[51] Put another way, if 99.999993% of instances of war have been immoral in the past, is it *worth* spending your time assessing whether the *next* instance is immoral? (Note that "assessing" a war means relying mainly on information your government provides, which past history also suggests is mainly lies.) A point I raised at the start is whether the existence of only one agreed-upon case of just war is not in fact a strong argument *against* war.

A relevant factor here is *why* there have been so few just wars in the past. Perhaps there is some explanation of this that would suggest why there might be *more* just wars in the future, thus affirming the rationality of assessing each new one as they come along. But just war theorists like McMahan provide no explanation for the striking absence of just wars in the past; indeed, they provide no account of why unjust wars have occurred because they provide no account of why wars occur, period. Explanations that appeal to, say, inevitable human frailties (self-interest, etc.) would seem to imply that the cavalcade of injustice will just continue. If, in fact, there has been only one just war in the past—World War II—a more reasonable explanation would seem to be that it was a purely *random* event, hence not the sort of thing one bases a policy on. By contrast, the pacifist *does* have an explanation for the paucity of past just wars, which refers to the logic of the war system.

Another way to frame things is why there is so little *agreement* about which wars have been just. This is posed by Saba Bazargan in his article,

50 McMahan, Jeff. *Killing in War*. Oxford University Press, 2009, p. 120.

51 Mack, Eric. "Astronomers Just Updated the Odds a Massive Asteroid Will Hit Earth in The Future." *Forbes*, Mar 29, 2022, www.forbes.com/sites/ericmack/2022/03/29/astronomers-just-updated-the-odds-a-massive-asteroid-will-hit-earth-in-the-future/?sh=538be2ec3a9c.

"Morally Heterogeneous Wars."[52] His concern is that, whether or not there are just wars, if we cannot be confident that any particular war is just then we are obliged to err on the side of caution—and refuse to fight in it. Hence, the problem of identifying just wars threatens to "lure us," in his words, into what he calls "epistemic-based contingent pacifism." Bazargan's response begins by noting that wars have a multiplicity of aims. He argues that the reason why there is such disagreement is that people try to aggregate these various aims together, to achieve some overall appraisal of a war's justice, whereas they should only focus on what he calls the "ultimate aim" of a war. But why should we expect any more agreement on "ultimate aims" than on aims generally? One could argue that in actuality the disagreements only become *worse* when one asks about "ultimate aims." States and their governments are complex, contested institutions which, as Clausewitz stressed, talk of their having "aims" only obscures. The complete lack of agreement over wars over time is not a matter of the framework people bring to war but of the political—hence inherently *contestable*—nature of war itself.

[b]/ Is It Appropriate?

A second question is whether assessing each new war as it comes along is *appropriate*.

The problem I have in mind involves the larger issue of the *individuation* of wars. Consider what Americans now call the "Indian Wars." Between 1500 and 1900, there were about 150 of them in North America. They increased dramatically in the second half of the 19th century. A single decade, the 1870s, saw the *Comanche Campaign* (1867–1875), the *Red River Rebellion* (1869–1870), the *Modoc War* (1872–1873), the *Great Sioux War* (1876–1877), the *Buffalo Hunters' War* (1876–1877), the *Nez Perce War* (1877), the *Bannock War* (1878), the *Cheyenne War* (1878–1879), the *Sheepeater Indian War* (1879), the *White River War* (1879), and *Victorio's War* (1879–1880).

Now suppose someone insisted that open-mindedness required assessing each of these wars separately and remaining open to the possibility of a "just" Indian war in the future. I would take this to be wrongheaded, and its wrong-headedness would involve treating these wars as discrete events. It is too specific as to both the contending parties and the causes of their conflict. Taken alone, the Sheepeater War was a conflict between the 1st Cavalry Division of the US Army and the Shoshone/Tukedaka occasioned by the alleged killing of two ranchers. But viewed in its larger—and

52 Bazargan, Saba. "Morally Heterogeneous Wars." *Philosophia*, 41, 2013, pp. 959–975.

proper—context, it was an instance of the ongoing conflict between Euro-Americans on the one hand and Indigenous peoples on the other caused by the former's incursions into native lands.

All of which is to say that these "wars" were not discrete events but aspects of a larger system called imperialism.

The point bears directly on war today insofar as official thinking is now comparing America's "long wars" to the "Indian wars" of the past. But we should not conclude that assessing wars one at a time is always a case of misplaced specificity. Whether wars can/should be individuated is itself a reflection of the larger system of which they are a part. There was "war" in feudal times but the feudal system did not involve "wars." In the post-feudal war system, wars were individuated in Europe but not outside of Europe—so the suggestion of assessing "each new war as it came" along never made sense there. It only made sense in Europe as part of the assessment of the larger war system.

[c]/ Is It Offensive?

A final question is whether assessing wars on a case-by-case basis might be *offensive*.

Consider this parallel. In the United States, there were 6,000 lynching from colonial times to the 1970s. Suppose someone said that open-mindedness required assessing each one on a case-by-case basis, because a "justified" lynching is certainly possible. I suspect that many would find this offensive, and not just those whose heritage had been touched by lynching. Indeed, I suspect that Native Americans would be offended by the suggestion that each Indian war be assessed on a case-by-case basis, given the possibility of a just one. And I doubt that their feelings would be mollified by someone picking out, among those thousands of lynching/hundreds of Indian wars a single case of a "justified" one.

[3]/ War as a Practice

Let me conclude with some words on how the disagreement between un-conditional and conditional pacifism reflects different conceptions of war as a practice.

The root idea of war as a social practice is that it is a social creation. What this means is, first of all, that war is not a biological necessity like eating or sex, but rather it is a social artifact—typically likened in this regard to dueling. But there are two different ways we might think of the notion of "practice" here.

One picture of war as a practice is what I call a *serial* conception of war. For this, "war" denotes a general type of which individual "wars" are tokens, sharing certain common features that mark them as criminal acts in the pacifist's eyes. Various proposals have been made for what these common features are. For Thomas Hobbes, "War is . . . an act of force to compel our opponent to do our will," for Quincy Wright, war is "the use of proportionate and discriminate armed force by legitimate public authority in order to secure certain worthy public goods," and so on. To say that war is a practice in this sense is to say that it is something that people/ groups "make a practice" of doing. Talk of "war," then, is talk of the ensemble of individual instances of war, with no particular assumptions about why they occur—or may continue to recur.

The other picture of war as a practice is what I call a *structural* one. For it, wars are not tokens of a more general type but aspects of a deeper institution. That institution explains why particular wars occur and why they will continue to recur until the institution is fundamentally changed. These pictures are not mutually exclusive, but adopting one over the other has significant consequences for the appraisal of the practice at hand.

Let me illustrate this by considering a contemporary issue where the difference manifests itself.

Consider the condemnation of rape, and the question of how rape is conceived. There is agreement that rape is a practice. But some people construe this in the serial sense; for them, talk of "rape" is talk of the ensemble of individual instances of rape all sharing certain features, like sexual violence. To view it as a practice is to view it as something that people, specifically men, habitually engage in; but there are no particular assumptions about *why* they engage in it. Others construe the practice of rape in a structural sense. For them, rapes are not just individual actions sharing common features; they are aspects of a larger institution where men's violence in rape is rooted in facts of male dominance. Or consider the practice of lynching. At one time, people had a serial conception of lynching: it was something that whites habitually engaged in against blacks, connected only by common features. Now, it is conceived as an institution, whose instances are aspects of a deeper and ongoing structure of white supremacy that lynching both manifests and sustains.

It seems to me that just war theorists and provisional pacifists adopt a serial conception of war, which explains why an unconditional condemnation of it is so problematic for them. For a serial conception, unconditional condemnation is a matter of how *broad* the condemnation is. It is a condemnation of each and every instance of war—which is challenged, then, by counter instances that do not apparently fit the criticism. For a structural conception, though, unconditional condemnation is a matter of how *deep* the condemnation is. Instances are condemned as aspects of a deeper practice, and the existence of counter instances may only show how that practice occasionally generates "good" instances to justify itself. The attraction of the serial conception is odd given the analogy often invoked with dueling, since clearly, when we talk of dueling, we are not referring to each individual duel in the aggregate but to the deeper institution of which individual duels were expressions.

At bottom, I think we return to an issue raised by Clausewitz. To think of the social practice of war as the ensemble of individual instances is to think of war as war making, hence to think of it for Clausewitz more superficially. To think of those instances as expressions of a deeper structure is to focus on war building as underlying war making. Which picture of social practice best fits war? I do not think there is an *a priori* answer to this, though our initial hunch will guide inquiry. Those who view rape or lynching as manifestations of a deeper social structure face the challenge of articulating what that structure is and why it manifests itself as such. Note that the appeal to social structure implies that any *solution* to rape or lynching must address that structure and not its surface manifestations. Accordingly, the claim that the social practice of war must be understood structurally faces the challenge of articulating what that structure is and how it manifests itself. I have already given an outline of this, and now will say more about it in the next chapter.

Chapter 4

The Dynamic of the War System

This chapter further develops the account of the war system by providing a thumbnail sketch of how it has historically unfolded over the last 500+ years. To say that it has a dimension of *1066 and All That* would be an understatement, but I have already voiced my excuse/apology in Chapter 1. We already carry thumbnail sketches of modernity in our heads, the standard ones being stories of progress in which war plays no important role at all. A starting point for challenging them is considering what an alternative thumbnail sketch might look like. My discussion is divided into three parts. I first consider the origins of the war system in states, empires, and their relation to each other. I then describe the stages of the war system, including the rise of the nation. I conclude with some of the challenges this account poses for reconceiving politics today.

[I]/ Medieval Frameworks and Modern Cycles

[1]/ The Feudal Background

The feudalism from which Western modernity emerged was a world of all-inclusive low-intensity conflict that blurred any distinction between war and peace. The stability of the Roman Empire, that is, its ability to maintain and govern large territories, had rested on a unique configuration of power consisting of a strong imperial state, or *imperium*, and strong private property, or *dominium*—clearly distinct from one another, in contrast with other empires, but strongly allied with each other via the type of contractual relation at the heart of Roman law. The collapse of the Roman Empire saw the collapse of *imperium* into *dominium*, in which heretofore centralized political power was dispersed outward and downward.[1]

---

1 Wood, Ellen Meiksins. *Liberty and Property: A Social History of Western Political Thought from Renaissance to Enlightenment.* Verso, 2012, pp. 4–7; see also Anderson, Perry. *Passages from Antiquity to Feudalism.* Verso, 1974.

DOI: 10.4324/9781032686189-4

What eventually emerged from this was the system we know as feudalism. As John Ferejohn and Francis McCall Rosenbluth write in their *Forged Through Fire: War, Peace, and the Democratic Bargain*, feudalism was basically "a protection racket by which the peasantry, virtually the only part of society creating value at that time, paid lords not only to protect them from other lords but also not to destroy them themselves."[2] Specifically, lords granted property rights/"fiefs" to vassals in return for their support in war/"fealty" when needed; vassals extracted wealth from the peasantry, especially for such implements of war as horses, weapons, and armor, alongside the taxes that lords imposed on peasants. All of this was justified as providing "security" to the peasantry—from the wars that lords and their vassals constantly waged against each other. Hence, the feudal system of alleged reciprocal relationships only served to obscure how the entire system rested on an ongoing act of extortion.

Feudalism's tangled web of rights and obligations gave rise to endless disputes which, in the absence of a coherent legal system, were ultimately resolved by battle—or what later came to be called "private wars."

Two different ethics of war sought to constrain feudalism's violence. They continue to define the frameworks employed today.

One was a more *reserved* ethic that traced its origins to Augustine's attempt to reconcile Roman/Stoic legalistic justifications of war with the Christian Gospels' skepticism of it. For it, war was a mournful enterprise, a sad necessity in a sinful world needed to restore a peace that would shelter the innocent. But war was never good or righteous in itself. It had to be waged without pride or malice. Since all taking of human life was sinful, it had to be atoned for afterward, however much justified. From this came our notion of war as law enforcement, in this case overseen by a common church. But it departed from later just war thinking in two significant ways. War was not about "self-defense" (this was the nod to Jesus' "turn the other cheek"); it was about helping others. And just war principles primarily assessed actions after the fact, for considerations of penance; they were not conceived as directing decisions to wage war because war was not regarded as an "instrument" of "policy," as it would later become.[3] This reserved approach was later augmented by the face-to-face ethic of combat known as "chivalry." It yoked violence to gender via an ethic of "manly" decorum that has proved remarkably enduring in the ideology of warfare and popular culture generally.

---

2 Ferejohn, John and Rosenbluth, Francis McCall. *Forged Through Fire: War, Peace, and the Democratic Bargain*. Liveright, 2016, p. 88.
3 Yoder, John Howard Yoder. *Christian Attitudes to War, Peace, and Revolution*. Edited by T. J. Koontz and A. Alexis Baker. Brazos Press, 2009, pp. 84–85.

The other was a more *aggressive* ethic most dramatically evidenced in the Crusades—of war as holy war.[4] With it, the picture of war moved from the mournful and tentative to the celebratory and certain. Ever since, the reserved and aggressive strands have vied for the soul of "just war" thinking. Does it enlist morality to counsel restraint, or weaponize morality to endorse aggression?

The Crusades were significant on several levels. Politically, they were an attempt by the church in Rome to consolidate its dominance—and constrain conflict within its realm—by mobilizing Christians to fight against a common enemy: Islam. The fight was soon extended to Jews, heresies, and the Eastern Church when it got in the way. It was Europe's first experience with mass popular mobilization (the equivalent of one million people today) that revived the project of empire in Europe and in so doing gave birth to the very notion of "Europe" as a distinct political identity. Theoretically, war assumed a new political function of binding people together by identifying a common enemy to kill—with major implications for how war was conceived. The reserved/law enforcement ethic had seen war as something permitted but not obligatory, whose sacrifices were regrettable, and which aimed at welcoming the adversary back to the community. Now the aggressive/holy war ethic saw war as something mandatory, whose sacrifices were honorable, and whose aim was the very survival of the faith by destroying its enemies. Moral principles played an exhortatory function, rousing believers to action.

These two pictures of war drew on a distinction from Greek and Roman times between two types of war, on that continues to inform thinking today.

Greek and Roman practice drew a sharp distinction between wars against other "civilized" peoples and wars against "uncivilized" peoples— that is, "barbarians" or "savages." The first involved what I shall term war, proper; the second what I shall term hostilities. War, proper (between "civilized" peoples) is a discrete act, bound by rules, with a finite beginning and end, and akin to a civil strife insofar as the aim is ultimately reconciliation. By contrast, hostilities (with "barbarians"/"savages") are an ongoing condition, bound by no rules, that only ends with the destruction of one of the parties. (Its impulse, then, is genocidal.) These have been identified with different notions of "self-defense." In war proper, self-defense is a discrete act—repelling attack. In hostilities, self-defense is an ongoing project—securing survival. Rome's imperial conflicts involved

4 Bainton, Roland. *Christian Attitudes Toward War and Peace: A Historical Survey and Critical Re-Evaluation.* Abingdon Press, 1960.

ever-expanding hostilities justified by the needs for survival, which Michael Mann had identified as a rationalization of all empires.

[2]/ Extrusion and Implosion, Paradox and Resistance

The emergence of modernity from feudalism involved a return to the separation of the political and economic realms that had been blurred in feudalism. Rome's distinction between *imperium* and *dominium* and the relation between them reemerged in the two forms of power identified in the previous chapter: states/political power and capital/economic power. The system that resulted was a new, more expansive and more powerful form of protection racket.

It came to be defined by a distinct dynamic.

Champions of the state portray it as "taming" feudalism's violence. What they have in mind is its ending feudalism's all-inclusive low-intensity conflict that blurred any distinction between war and peace. But the state did not end violence so much as push it out to its borders, so that "private" wars between individuals became "public" wars between states. As F.H. Hinsley notes, "The first effect of the state's growing monopoly of armed force within society was an increase in the frequency of interstate war."[5] "Peace" became distinguished from "war" as what existed within states and what existed when the conflicts between states were suspended. And the dynamic did not end there. In time, the consolidation of states into the European state system pushed violence out further—to the frontiers of European empires. So eventually violence within states tended to be tamed, violence between states tended to be contained (via the adoption of just war theory's reserved ethic, defining European conflicts as wars, proper), leaving the violence of empires between the so-called civilized and the so-called non-civilized worlds as completely unrestrained (via the adoption of just war theory's aggressive ethic, defining European/non-European conflicts as hostilities).

Let us call this feature of the war system its *logic of extrusion*. Violence is pushed from the center to the periphery, where violence at the center is the orderly constrained enterprise of war, proper, while violence at the periphery is the disorderly unconstrained enterprise of hostilities.

But a further feature of the dynamic is how this logic has periodically— and catastrophically—broken down. Let us call this its *logic of implosion*.

---

5 Hinsley, F. H. *Nationalism and the International System*. Hodder and Stoughton, 1973, p. 92. Cited in Alexandra, Andrew. "Political Pacifism." *Social Theory and Practice*, 29(4), 2003, pp. 589–606.

Modernity has (so far) witnessed three "hyper wars": the Thirty Years War, the Napoleonic Wars, and World Wars I and II. Each lasted about three decades. Each involved the attempt by one state or group of states to conquer all of Europe so as to create, as it were, an empire in Europe. In each case, the attempt at European empire was unsuccessful. But in each case, it led to a lesser form of dominance that scholars term *hegemony*.[6] By this, they mean that one power is dominant enough to impose its rules on the political, economic, and even cultural areas. The Thirty Years War resulted in the hegemony of the United Provinces, reflected in the Treaty of Westphalia. The Napoleonic Wars resulted in British hegemony, reflected in the Concert of Europe. World Wars I–II resulted in the hegemony of the United States, reflected in the United Nations/Bretton Woods, whose fate today remains uncertain.

Each of these conflicts was informed by factors outside of Europe. The Thirty Years War was driven by the riches of imperialism; the Napoleonic Wars and World Wars I and II were provoked by conflicts over imperialism. Hence, violence on the periphery returned to the center as Europeans proceeded to treat each other with the same savagery they had otherwise inflicted on the non-European world.[7] Finally, each conflict involved and promoted political and military developments that expanded the means of destruction. For example, World Wars I and II were both the result of industrialized warfare of a type that had been developing since the 1840s and the consummation of such warfare in the development of nuclear weapons that served to define the post-World War II era.

The final feature of this dynamic is its logic of *contradiction*.

The story of war has been a story of paradox. Nowhere was this more true than in the 20th century, whose warfare–welfare states made it the century of both the graveyard and the supermarket. It is estimated that between 1500 and 1800 average per capita global domestic product increased by little more than 50%. Between 1870 and 1998 it increased by a factor of more than 650%. Put differently, compound annual growth rate was nearly 13 times higher between 1870 and 1998 than it was between 1500 and 1870. I have stressed the paradox of martial arrangements and liberal rights but it will be important to keep this in perspective. Initially, these liberal/democratic arrangements were holdovers from the medieval era, especially as they had been nourished in the towns and city-states

---

6 Wallerstein, Immanuel. "The Three Instances of Hegemony in the History of the Capitalist World-Economy." *International Journal of Comparative Sociology*, 24, 1983, pp. 100–108; Arrighi, Giovanni. "The Three Hegemonies of Historical Capitalism." *Review* (Fernand Braudel Center), 13(3), 1990, pp. 365–408.

7 Mazower, Mark. *Dark Continent: Europe's Twentieth Century.* Vintage, 2000.

outside of feudal arrangements. Their endurance resulted from the bargain of rulers accepting their continued authority in return for funds to wage their wars. But other bargains have been more straightforward. For example, every expansion of the voting franchise in the United States has resulted from war. In the 20th century, women were granted the vote for their support of World War I and 18-year-olds were granted the vote for their participation in the Vietnam War.

But the role of these liberal/democratic arrangements and their ideologies has been deeply *ambivalent*. Sometimes, they have served to enable militarism, for example, when they have been enlisted in Europe's holy wars against non-European peoples—wars for "civilization" like Europe's conquests in Africa or more recently America's "War on Terror." At other times liberal/democratic arrangements have sought to constrain war, and their ideologies have been deeply critical of war. This has been especially true in times of crisis. The 18th-century crisis saw Rousseau's challenge to the whole notion of the state and Diderot's sharp attacks on European imperialism—both of which laid the groundwork for political pacifism's later coalescence. The 20th century saw the militarization of Europe's dominant ideologies, liberalism, democracy, and socialism. But it also engendered resistance from within, such that political pacifism can be understood as extending the peace-oriented dimensions of those ideologies. A theme of Jonathan Schell's *Unconquerable World* is how the history of the war system has also involved "a less noticed, parallel history of nonviolent power." In that sense, to borrow a phrase from Marxism, we might speak of the war system as "digging its own grave."

This only pertains to the Euro-North American world. Outside Europe, imperialism has destroyed local communities and degraded local political structures so that they have received all the burdens of war and none of the benefits, as it were. But this has been met with fierce resistance. The European thinkers critical of imperialism were deeply influenced by the movements and voices of those non-European resisting it. In the 20th century, such resistance led to the deepest challenges to the war system in two respects. Practically, anti-imperialism was the incubator of nonviolent forms of resistance, which were then adopted in the Euro-North American world itself. Theoretically, it has been non-European voices such as those referenced in Chapter 2 that have most identified Western modernity as a system of war and have most called for political/cultural revolutions of peace to replace it.

[II]/ The Global System

The dawn of modernity is commonly dated around 1500. The political institutions of modernity were forged in the almost constant warfare before and after that time. At the heart of these conflicts was the question of which political form would prevail: states or empires? This ended around 1650 when what I shall term the Westphalian Settlement laid the framework for the modern state/global system.

Let me begin with some words on states and empires separately, leading to how they merged with the emergence of the global system.

[1]/ The Emergence of the State

The standard story of the state is deeply indebted to Michael Roberts' thesis on the "military revolution" of the 16th–17th centuries.[8] The details of his thesis have been contested, but his larger claims about the relation between war and the state and its importance for understanding the ferocity and scale of 20th-century war remain secure. Basically, technological and organizational developments in war dramatically increased the size of infantry-based armies. The introduction of gunpowder and with it changes in artillery and infantry formations led to more offensive warfare and much bigger armies that required stricter discipline and organization for their employment. For example, Spain's army grew from tens of thousands in 1500 to 300,000 regulars and 500,000 militia by 1650. It all meant a dramatic increase in financial and organizational demands that only a new political entity—the state—could manage. To meet those demands, states were driven to consolidate their domestic authority to mobilize the human and material resources needed.

This political form was not entirely novel. It was indebted to the classical Greek city-state and its emphasis on the territorial organization of states at the expense of alternative political associations. It was first revived in the city-states of Italy between the 13th and 15th centuries and then spread to Europe generally where it marked the death knell of both the localisms of feudalism and the universalisms of the church. Greece also foreshadowed how war building could be an engine of democracy. Full-fledged democracy in Athens was born when Athenians were granted political power in return for manning the hoplites necessary to repel Spartan invasion. This was a limited democracy, as hoplite participation required armor and weapons that only landholders could afford. It was expanded when the expansion of the navy required that poor citizens be given political power

---

8 Parker, Geoffrey. "The Military Revolution, 1560–1660—A Myth?" *Essays in Swedish History.* Edited by Michael Roberts. J. Wyatt Books, 1967, pp. 195–196, 218; Downing, B. *The Military Revolution and Political Change: Origins of Democracy and Autocracy in Early Modern Europe.* Princeton University Press, 1992; Black, Jeremy. *A Military Revolution? Military Change and European Society, 1550–1800,* Red Globe Press, 1991. Ertman, Thomas. *The Birth of Leviathan.* Cambridge University Press, 1997; Van Creveld, Martin. *The Rise and Decline of the State.* Cambridge University Press, 1999.

in return for manning the triremes. (Aristotle termed the constitution of Athens a "democracy based on triremes.")[9]

The powers of the modern state were magnified by a crucial *abstraction*. This was the act of reification whereby the attributes of persons, that is, princes or "sovereigns," were transposed onto an abstract entity, the "state" and its "sovereignty."[10] As Philip Bobbitt writes, "All the significant legal characteristics of the State—legitimacy, personality, continuity, integrity, and, most importantly, sovereignty—date from the moment at which these human traits, the constituents of human identity, were transposed to the State itself." Bobbitt describes how this abstraction contributed to the state's ability to mobilize the powers of war making, while this same abstractness also gave rise in Bobbitt's words to sovereignty's "unique problem, its problematic relation to the elusive status of legitimacy."[11] Rousseau would suggest that its elusive status rendered the "state" permanently insecure, if not neurotic, needing to confirm its abstract existence by attaching itself to that most material/bodily human endeavor, armed conflict. Later critics would raise the same point about another abstract—and elusive—entity, the "nation."

The alliance of state and capital noted in the previous chapter took nascent form in the alliance of monarchs and towns. The role of the town was crucial, as scholars like Lewis Mumford have stressed. Their importance lay in the extent to which they had not been subsumed under the hierarchical relations of feudalism and its lord/vassal/peasant relations. They were "zones of freedom" making them sites of market transactions, hence incubators of emerging capitalism, as well as sites of political experimentation in terms of community empowerment. This enabled monarchs to triumph over the nobility to achieve the centralized control that was the hallmark of the state. I have noted that this proceeded both internally and externally.

The external process, of rulers seeking the realms of other rulers, meant an era of almost constant strife between states. Whatever their outcome, these "foreign" wars were spurs to "domestic" consolidation. The One Hundred Years War (1337–1453) prompted political consolidation and proto-nationalism in both England and France. The English state achieved maturity under the Tudors via the War of the Roses (1455–1487), a direct consequence of the One Hundred Years War. Around the same time, the

---

9 See Hale, John R. *Lords of the Sea: The Epic Story of the Athenian Navy and the Birth of Democracy*. Penguin, 2009. Ferejohn, John and McCall, Francis Rosenbluth, *Op. cit.* Chapter Two.
10 Marx held that the triumph of capitalism involved a similar abstraction, whereby the material properties of commodities were transposed into an abstract entity, "value."
11 Bobbitt, P. *The Shield of Achilles. Op. cit.*, pp. 82–83.

Italian Wars (also known as the Hapsburg-Valois Wars, occurring off and on from 1494 to 1599) furthered the consolidation of the French state. The Spanish state was consolidated by the "Reconquista" expulsion of Islam that concluded at the end of the 15th century. This was, it might be noted, the last—and only successful—Crusade, which would bequeath a crusading legacy to Spain.

The internal process, whereby monarchs achieved dominance within their existing realms, involved both negotiation with and compulsion against competing powers. The balance between these factors was of decisive importance for the type of state that prevailed, specifically for the extent of its liberal/democratic arrangements. Such arrangements were not a blessing from the new state, as statist ideologies would have it. Rather, they were holdovers from the participatory institutions of previous times that endured due to the logic of state formation: where *resistance* was strong to the state and its war making endeavors, the resulting state granted individual and group rights (to the privileged, primarily); where it was weak, those rights were swept away.

Hence, the European states that emerged assumed different political forms depending on the relative roles of negotiation and compulsion. This would cast its shadow over the subsequent history of Europe.

The most successful political form to emerge involved what has been termed the *constitutional* path to state building. This was exemplified by England, and its effectiveness at war building explains how Britain could go from a minor political power on a marginal set of islands off the European coast in the 16th century to a major power in the 17th and 18th centuries to a dominant power in the 19th and 20th centuries. Its version of the war contract was between the monarchy and a middle class of merchants and city dwellers against the landed power of the nobility. The power of the middle class in Parliament's House of Commons meant that it could provide revenue for the monarch's wars in return for the crown's continuing to respect the constitutional arrangements of the medieval era. The upshot was to mute the bureaucratic and centralizing impact of war. The collapse of this relationship in the English Civil War was when the monarch's frustration at Parliament's resistance to funding its wars led him to assert royal absolutism, an act that proved fatal to him and led to the arrangements of constitutional monarchy.

Similar bargains created the even more liberal/democratic arrangements in the Netherlands, Switzerland, and Sweden. This has been termed the *coalitional* path, where the resistance of non-state constituencies meant that the representative assemblies of medieval origin survived almost fully intact. But the principal contrast was with the *continental* path, exemplified by France and Spain. Here, the relative weakness of other constituencies meant that the monarchy had a free hand in appropriating resources. It felt no compulsion to bargain with other constituencies, hence no need to respect traditional institutions or local privileges and liberties. The French government raised more money when the traditional Estates were not in session, the Spanish monarch refused to summon them for decades at a time. The French Revolution resulted when the monarch was no longer able to pay for war this way so was compelled to engage traditional representative institutions, an act that proved fatal to him and led to more republican arrangements.

"Given the imperatives of the Military Revolution," Bruce Porter writes in *War and the Rise of the State*, "the real wonder of the age was not the rise of the state, but the survival of constitutional and representative systems in England, Sweden, Switzerland, and the Dutch Republic."[12]

[2]/ The Emergence of Empire

As previously noted, empire preceded the rise of the state. It too was a legacy of the Crusades in two respects. It was the Crusades and the links they established with the East that nourished the development of the Italian city-states economically and culturally. Venice's emergence as a maritime power led to its creating a proto-empire in the Mediterranean that was the first sea-born empire since Rome. In time, though, it was the Islamic expulsion of Christian forces from the holy lands that ended the transcontinental paths to the Far East and compelled the search for alternative sea routes—and with it, the emergence of European imperialism.

If the state originated in the alliance of monarchs and towns, modern empires originated in the alliance of monarchs and city-states. The latter included not just Venice but other Italian city-states, Genoa especially. Together, they were the incubator for the state system and the originators of the practices of diplomacy as part of the first "balance of power" arrangements. Both were empowered by the fact that, along with its links with the East, Italy's geographical position insulated it from feudalism and later monarchical conflicts, at least for a time.

The first imperial ventures were Portuguese, reflecting Portugal's geographical isolation from the rest of Europe that compelled it to the seas. It expanded southward along the African coast and eventually to the Indian Ocean, exploiting developments in naval technology from the Military Revolution. The real leap was the entrance of Spain in alliance with the financial interests of Italy. It was the Genoise Christopher Columbus who began the Spanish incursions into the Americas, which many date as the true beginning of the modern world. At the end of the 16th century,

---

12 Porter, Bruce. *War and the Rise of the State*. Free Press, 2002, p. 103.

the English and the Dutch began to challenge Iberian monopolies on trade. Their particular focus was the lucrative spice trade in the East Indies. Their victory over Portugal led to the Anglo-Dutch Wars of the 17th century that only ended when the Dutch William of Orange was offered the English throne in 1688's "Glorious Revolution."

The English initially trailed others in the enterprise of empire. Their first effort was not in Asia or the Americas but in Ireland, starting with the 16th-century colonization of Ulster by English Protestants. English "plantations" in Ireland would remain the model for colonization elsewhere. Its take-off was empowered as I have noted by the joint stock companies pioneered by the English and the Dutch. The focus of the so-called First British Empire was the West Indies, where the "Sugar Revolution" of the mid-17th century made the Caribbean its most important and lucrative colonies. The enormous wealth generated there financed the development of non-plantation colonies in North America—and the ongoing conflicts with France that this engendered.

The upshot was the centrality of slavery to British economics and politics. Large sugarcane plantations initially used native and white indentured labor but quickly turned to the importation of slaves. British ships carried approximately 3.5 million Africans across the Atlantic. By 1780, the percentage of slaves in the population of the British Caribbean rose to around 80%, and in the American colonies to around 40% (mainly in the southern colonies). Britain was transformed, as its port cities of Bristol, Liverpool, and London were responsible for the bulk of British slave trading. The "Second British Empire" of the 19th century revolved around the cotton harvested by slaves in America along with the wealth extracted by British control of India.

Empire initially adopted the ethic of holy war, except that now the Crusading impulse was diverted from the East to the West. It was this ethic that underlay the enterprise's exceptional cruelty. If the growth of states within Europe meant the dispossession of local populations, the growth of empire outside Europe meant their decimation. But this story also had its paradoxes. The initial incursions into the Americas were flatly genocidal. This inspired a critique from within the Catholic Church that asserted the rights of Indigenous peoples as equal human beings with souls. This is generally regarded as the first expression of human rights doctrines that would come to play such a major role later. But its impact at the time was deeply ambiguous. The condemnation of genocide did not extend to a condemnation of slavery; rather, a justification for the latter was found within the moral heritage of Europe itself—its just war theory. Enslavement was portrayed as a humanitarian alternative to being killed as a consequence of defeat in war. This was later supplemented by figures like Vitoria with paternalist arguments for governing the Indigenous "for their own benefit" by putting an end to their "sacrilegious rights" of human sacrifice, cannibalism, etc. Nineteenth-century empire replaced "Christianizing" with "civilizing" as its overarching ideology, replacing "saving souls" with "progress" as the justification for imposing its arrangements on others. To this, it fatally added racism, which was not a prominent feature until the rise of social Darwinism in the later 19th century.

Alfred McCoy has argued that this gave rise to a further sense of "sovereignty" whose legacy remained firm well into the 20th century.[13] As noted in the last chapter, state sovereignty, which prevailed in Europe, meant concentrated power over a specific area defined by borders. Imperial sovereignty extended outside of Europe. It meant the right of European states to extend their power over vague areas defined by frontiers; and those areas could be vague indeed, sometimes extending over whole continents or oceans. In that sense, it was a more expansive form of sovereignty, though the power claimed over other areas was itself often diffuse. This inevitably led to the ongoing conflicts between empires outside of Europe which ultimately returned to transform Europe.

[3]/ The Westphalian Settlement and State–Empires

The conflicts of nascent states and empires led to the first crisis of modernity generally identified with Thirty Years War (1618–1648), though this was itself the combination of struggles that began with the Dutch revolt against Spain against Spain (1568–1648). The whole era was an incoherent mishmash, as anyone who wrestles to comprehend the "Thirty Years War" learns. Its brutality and futility—and apparent endlessness—were dramatized in Bertolt Brecht's *Mother Courage*, in which a wily canteen woman comes to make a living off the pointless conflict only to see her children perish in the war.

The chaos of the wars posed serious challenges to the legitimacy of war itself and the entities that waged it. It generated widespread popular revolt against rulers who had created a world so evidently absurd. It witnessed the first flourishing of a non-religious pacifism. So legitimizing war meant *reconceiving* it—articulating a conceptual structure that imposed some coherence on it. How this was done is a long story involving, among other things, scuttling feudal notions of war as a private affair for the early modern conception of "public war." Theorists of the modern state like Machiavelli and Cardinal Richelieu fashioned an alternative to war as a sacerdotal melee. War was now an "instrument" of "statecraft," fought by

---

13 McCoy, Alfred. *To Govern the Globe: World Orders and Catastrophic Change*. Haymarket Books, 2021, pp. 33ff.

political leaders for defined "reasons of state." With this, it bears stressing, the whole conception of "just war principles" changed. In the Middle Ages, such principles functioned as a largely after-the-fact affair: they were there to appraise actions after they happened, largely for the purposes of penance, with no assumption that they could direct wars as they were conducted. With the modern instrumentalist conception of war, "just war principles" came to be conceived as directives to be applied by leaders before the fact and during the fact.

Ultimately, the response to this legitimation crisis is what I term the *Westphalian Settlement*, with the Peace of Westphalia (1648) standing for a cluster of commitments and beliefs that coalesced over time. It would frame European/global politics for the next three centuries.

At its heart was the fashioning of a new political form: the *state-empire*. Basically, it resolved the tension between state and empire by assigning them to different parts of the globe. One may think of it as extending the war contract from one *within* states, driving their conflicts with each other, to one *between* states, defining their wars with the rest of the globe. This agreement between states gave birth to "public international law" as something aimed at defining both relations within Europe and relations of Europe with the non-European world. "Sovereignty" no longer meant just priority within one's own realm, it now included equality with others equally sovereign (within Europe) and superiority over those lesser sovereign or non-sovereign (outside Europe).

The terms of the agreement were this:

Within Europe, it was agreed that there would be no single empire but rather a community of multiple states. Furthermore, it was agreed that Europe would not be ruled by a single faith but that matters of faith would be left to each state. In their relations with each other, then, states no longer appealed to a transcendent religion in the hope that peace would be secured by the triumph of that religion. Rather, states would be guided by "reasons of state" in the expectation that peace would result from the "balance of power" maintained by each pursuing its own interests. Note how this furthered the reification of the "state," as it was increasingly conceived a quasi-person, possessing its own agency, capable of acting on its own "reasons" in pursuit of its own "interests," hence worthy of being entrusted with respect for the European community of states. Outside Europe, though, matters were different. There, empire was the dominant political form, typically ruled by a single religious faith, where projection of power was the rule. Whatever political arrangements already existed, they were not regarded as possessing "interests" or capable of acting on "reasons," so were not regarded as candidates for the agreements that defined relations between European states. William McNeill has suggested that empire at home was prohibited so as to facilitate imperialism abroad;

in his words, imperialism was a "safety valve" that diffused domestic conflicts.[14] But it would periodically inflame such conflicts as well.

As the crisis of the 17th century had resulted from hyper-war, a major concern of the Westphalian Settlement was preventing this from happening again. It did this via the framework introduced earlier. Conflicts within Europe would be wars, proper, regulated by the more reserved side of just war theory now secularized in public international law by figures like Grotius. The main changes were twofold: whereas in its original form war was something regrettable requiring penance afterward, now war was something honorable as a service to the state; and whereas in its original form war was not about self-defense but defense of others, it was now about "security"—defense of the state as defense of oneself. By contrast, conflicts outside of Europe would be hostilities, licensed by the more aggressive holy war model that was eventually secularized by "civilization" replacing "Christianity" as the gift that would be imposed on others.

Finally, I see the 17th-century crisis as giving rise to Europe's two main political ideologies: liberalism and republicanism. As I construe them, both were critical responses to war's excesses and sought to be counsels of moderation. I shall return to them when I consider the critiques of war that have arisen from within the war system.

For liberalism, the excesses of war demonstrated the dangers of religious passion and its crusading impulses. Hence, the response to war lay in separating religion from politics to prevent domestic factions from crusading against each other internally (the source of civil conflict), and to prevent states from crusading against each other externally (the source of international conflict). This foreshadowed later liberalism's separating the private from the public and the good from the right so as to cleanse politics of metaphysical claims—as incentives to a type of holy war. But this vision had a dark side as well. If rationality was the key to peace, then war could be justified in the name of peace as the spreading of rationality to those less rational. This is essentially how liberalism lost its critical antiwar edge in the 19th century.

For republicanism, war's excesses were not the product of irrational passions but of concentrated predatory power. Hence, the solution in both the domestic and international realms lay in keeping power dispersed. From the start, then, republicanism was more concerned with the nature of political organization. As Daniel Deudney has noted, "republics" were often contrasted with "states" in their more absolutist forms; plus, republicans were skeptical of empires as both expressions of and enablers of illicit

---

14 McNeill, W. H. *The Rise of the West: A History of the Human Community*. University of Chicago Press, 1963, p. 312.

coagulated power.[15] The dark side of republicanism was how war could be justified as extending republican arrangements in the name of peace. There is also the complicated relation of republicanism to "democracy," which I shall turn to in my remarks on nationalism.

[III]/ The Stages of the War System

From the start, then, the role of the state system was to maintain the state's capacity to mobilize the means of destruction while ensuring that it did not degenerate into the chaos that characterized the crisis of the 17th century. After the Westphalian Settlement, the story can be told in three phases, each inaugurated by the crisis of the previous phase, with the third phase being the one we are in now whose outcome remains to be determined.[16]

The *first* phase of the 17th–18th centuries following the Westphalian Settlement centered on *dynastic states* and their *mercantile empires*. War was the principal business of the state, but war in Europe—while common—was limited in character, while hostilities outside of Europe were unlimited.

This era was marked by several features:

- States and the wars they waged were largely *separated* from the rest of society. Populations generally had little involvement with or even awareness of the wars their countries fought.
- Wars were fought by standing armies that served the monarch's private purposes. These were ascribed to "reasons of state," but those reasons were the monarch's alone with little bearing on society as a whole.
- Standing armies were composed of mercenaries and professionals skilled in the use of the new firearms that were generally utilized for more defensive measures like sieges, and they fought for pay (the term "soldier" derives from "one who fights for pay").
- It followed that the principal challenge for the state was raising enough money to pay for war. Because society had little involvement with war, the tax base was limited so states had to rely on borrowing, which they did with varying success. Wars often ended when one or both sides ran out of money, and the system as a whole collapsed when the Seven Years War—which began as a conflict in the Americas that dragged all of Europe into it—brought the system as a whole to financial collapse.

---

15 Deudney, Daniel H. *Bounding Power: Republican Security Theory from the Polis to the Global Village*. Princeton University Press, 2007, p. 291.
16 The best short overview is Howard, Michael. *War in European History*. Oxford University Press, 2009.

- Critiques of the system from within Europe focused on the conflict between monarchs/standing armies and the enlightened interest of society as a whole, critiques from outside Europe dwelt on the predations of empire—that proceeded largely outside of European consciousness.

The *second* phase of the 19th–20th centuries following the Napoleonic Wars centered on *national states* and their *free trade empires*. Wars in Europe occurred less often, but they were now less limited in character—culminating in the unlimited total wars of the 20th century.

This era was marked by these features:

- War was now *integrated* into society politically, economically, and even culturally. Populations were increasingly engaged with the wars their countries fought—via the notion of the "nation."
- Wars were fought by conscripts/reserves that were motivated by shared ideological purpose rather than pay—for example, for the "values" the "nation" "stood for." They were generally utilized more offensively, with battle becoming the focus of conflict. (Over time, this led to the conception of war being decided by a "decisive battle.")
- Starting in the mid-19th century, these larger armies were capable of being rapidly mobilized by new modes of transportation like railroads and new modes of communication like the telegraph.
- The principal challenge for the state increasingly became one of mobilizing the populace to fight in wars, as well as pay the taxes for them. World War I ended with the disintegration of first the Russian and then the German war efforts, both among the troops and at home. This followed widespread mutinies among the Allies on the Western Front. Much of World War II's strategy was dictated by this concern, in Germany as much as anywhere else.
- The "nation" was the vehicle by which states mobilized for war, but the relation between the nation and the state was always complex—as each made claims to the "sovereignty" that was previously the monopoly of the state. Resistance to the system was often understood as resistance of the nation *to* the state, for example, of the "people" against the "war makers." As war required mobilization of the citizenry, protest took the form of resistance to conscription, war taxes, etc.

The *third* phase is what has followed World Wars I and II, and whose character is still in flux. Sir Michael Howard has termed post-Napoleonic War "nationalized war" and today's era "post-nationalized" war. Its features include the following:

- States and the wars they wage are now *alienated* from the rest of society in this sense: citizenries continue to identify with the wars their countries fight, but they refuse to participate in them—what I have termed the "Great Defection." Wars are financed by borrowing, as citizenries are also resistant to paying for them.

- Armies are composed of professionals and military-scientific elites. The former are "honored for their service", the latter are invisible to the public.
- The decisions to go to war are made by political/military elites, with little concern for input from the citizenry as the citizenry's involvement is no longer necessary.
- The development of nuclear weapons at the end of World War II was the consummation of 19th–20th-century industrialized warfare. They provide the capacity to inflict violence at an unprecedented scale and speed.
- The post-World War II era has been one of constant wars somewhere, but the disconnection of the citizenry from war has meant that for Europeans and North Americans war has been largely a "spectator sport" involving "a-symmetrical" conflicts, while the rest of the world has largely reverted to forms of low-intensity conflict.
- Since war no longer requires the citizenry's involvement, protest against war no longer involves the withholding of that involvement in draft refusal or tax resistance. Protest takes the form of popular demonstrations and the like, which wax and wane with citizens' attention. The Nuclear Freeze Movement of the 1980s was one of the largest mass movements in American history, but while the nuclear threat remains the issue has faded from popular politics.

[1]/ The Dynastic State: Deficits, Drill, and Desolation

Now for more detail on the dynastic state, and some words on the criticisms of it that emerged at the time.

Another term that scholars employ for the 17th–18th-century state is the *fiscal military state* to stress the fact that its principal arrangements were oriented to raising revenue for war.[17] Military costs generally accounted for 50% of state expenditures in peacetime and up to 90% in wartime. I term it the dynastic state to mark its rather tenuous relation to the rest of society, a disconnection that posed a significant challenge to the state's raising funds for war. States were headed by traditional ruling families who regarded the state's territory as akin to their private possessions and the

---

17 Brewer, John. *The Sinews of Power: War, Money and the English State, 1688–1783.* Harvard University Press, 1990. Bonney, Richard. *The Rise of the Fiscal State in Europe, 1200–1815.* Oxford University Press, 1999.

populace as akin to their tenants. Whatever bonds the populace felt for rulers were a matter of the national religions of the post-Reformation era. But generally, the populace's commitment to rulers was not much more than any tenant feels toward their landlord. As hirelings, soldiers had no particular affinities for the societies for which they fought; they and their officers often came from different countries. The British military that fought in the Americas was typical in the significant percentage of the troops that came from Germany and did not speak English.

The personalized nature of rule meant that wars were about dynastic succession or trivial issues of "prestige" that consumed royal cultures. Border disputes sometimes played a role, though European borders were basically fixed by the start of the 18th century. If these political issues dominated European wars, economic issues dominated wars outside of Europe. Prior to the 19th century, imperialism was driven by the quest for commodities that could not be gotten back home, or in the same quantity: precious metals, luxury items like sugar, etc. In this borderless world, competition easily led to conflict between Europeans, plus there was the ongoing threat of Indigenous peoples—who were invariably dragged into the conflict between Europeans.

The problem of paying for war explains why conflicts, though limited, dragged on. Both sides often had little interest in actually fighting given the expense of finding new soldiers. They avoided battles, and when they did the casualties were small. The Battle of Blenheim (1704), central to the War of Spanish Succession, saw around 10,000 soldiers killed. The Battle of Zorndorff (1758), a major conflict of the Seven Years War between Russia and Prussia and one of its bloodiest battles, saw around 25,000–30,000 killed. Contrast this with World War I's Battle of the Somme (1916) in which 300,000 soldiers were killed, or World War II's Battle of Stalingrad (1942), in which 1 million or more lost their lives. Furthermore, the rulers of dynastic states needed to keep their militaries intact since, prior to domestic police forces, militaries were the principal means of controlling people back home. So wars could proceed with little impact on domestic populations. In Lawrence Stearn's *Tristram Shandy*, the protagonist travels from England to France and back without realizing that the two countries are at war.

The financial mechanisms of the modern state arose to pay for war, and war has remained the principal influence on how states organize their finances.

The challenge was that rulers could not begin to pay for wars from current tax receipts, so they had to resort to borrowing. Even prior to the Westphalian Settlement, this had engendered a "financial revolution" in the development of centralized state instruments to finance wars. And it meant that success in war would go to those with the greatest capacity to borrow—which mainly meant who had the greatest capacity to maintain the trust of lenders (who were generally international in scope). The leader in this regard proved to be Britain, whose "Glorious Revolution" (1688) not only gave it a Dutch king but also engendered a Dutch mode of financing in the 1693 establishment of the Bank of England. Its genius was the invention of the government bond (or "consol") that enabled the state to float permanent public loans for war, and which was made possible by the alliance of the monarch with commercial interests in Parliament that inspired confidence that the loans would continue to be repaid. By contrast, the absolutist arrangements of the French monarchy meant that it lacked such confidence and hence was constantly constrained in its ability to borrow. Of the British system, Porter has noted, "Taxation by representation made it feasible to sustain higher levels of taxation and elicit greater national sacrifice in war than was possible through military-bureaucratic coercion."[18]

Finances, taxation, and the like are among those aspects of war building that are generally invisible. Their story is much less exciting than those of heroic deeds on the battlefield. But nothing is more important to the story of war and to the story of its larger political impact. The problem of paying for war, especially for debts already incurred, has been a major cause of modern revolutions. The after effects have been equally significant. War-induced inflation doomed the Weimar Republic and the concern that this experience not be repeated on a global level inspired the post-World War II financial arrangements of Bretton Woods.

Not surprisingly, finance became a focal point of 18th-century criticisms of the war system.

The late 18th century saw the dynastic state/mercantile empire system enter what would prove to be a terminal crisis. The American Revolution was one expression of this, exemplifying how the challenges of maintaining empire reverberated back to Europe itself. It was amidst this crisis that some of the most forceful critiques of the war system were advanced by Enlightenment thinkers.

A case in point was Kant's famous essay, "On Perpetual Peace" (1795).[19] One of its central and better-known themes was its condemnation of standing armies as exemplifying how the instruments of war could be employed for the most trivial of reasons and for no particular public purpose. In the same vein, Thomas Paine, Kant's contemporary, chastised dynastic military arrangements for meaning that wars could be fought "for

18 Porter, *Op. cit.*, p. 120.
19 Kant, Immanuel. *Perpetual Peace and Other Essays on Politics, History, and Morals.* Translated by Ted Humphrey, Hackett Publishing Co., 1983.

sport" by monarchs who were "pleased with any war, right or wrong, be it successful."[20] Before, though, rulers had been limited by costs. Now, this led to a further and lesser-known theme of Kant's critique: his condemnation of new forms of war financing which threatened to make the resources for war unlimited while hiding the true costs from the populace. The object of his critique was the "British system," in his words that "dangerous" practice that was "the ingenious invention of a commercial people in this century." Some will recognize a parallel here between Kant's critique of military arrangements and his critique of metaphysics. Just as in his view unconstrained metaphysical systems led philosophers on wild goose chases (which Kant blamed for introducing "civil discord" into the realm of reason), so too irresponsible unconstrained military/financial systems led societies on wild goose chases that Kant foresaw as ultimately a source of ruin.

Here is a remark from Kant's "What is Enlightenment?" (1784):

> All that is required for this enlightenment is freedom; and particularly the least harmful of all that may be called freedom, namely, the freedom for man to make public use of his reason in all matters. But I hear people clamor on all sides: Don't argue! The officer says: Don't argue, drill! The tax collector: Don't argue, pay! The pastor: Don't argue, believe! . . . Here we have restrictions on freedom everywhere.[21]

A further object of critical attention was the practice of military drill, which was taken to exemplify the noxious effects of military arrangements generally.[22] A number of practices resulted from the Military Revolution that we now identify with soldiering *per se*. The Dutch pioneered long hours of repeated drill as necessary to make soldiers recruited from the lowest ranks of society more obedient. Regularized marching taught them to move as a unit in prescribed patterns. The whole distinction between the "military" and the "civilian" realms arose to designate the differences between ordinary life and the routinized life that soldiers now entered, subject to a harshness of discipline that would not have been tolerated by soldiers in past times. Indeed, the transition to modern armies involved a dramatic loss of individualism generally. Sixteenth-century soldiers had prided themselves on the diversity of their raiments, typically gained by looting, with undergarments of silks and velvets. By the end of the 17th century, soldiers were wearing

---

20 Paine, Thomas. *The American* Crisis. 1776. CreateSpace Independent Publishing Platform, 2017, p. 25.
21 Kant, *Op. cit.*, p. 42.
22 McNeill, William H. *Keeping Together in Time: Dance and Drill in Human History.* Harvard University Press, 1995.

"uniforms" of a sort that identified them as figures of servile employment with restricted rights and liberties. All of this contributed to the conception of the military as an obedient "instrument" of state rule. Thus, Kant equated drill with intolerance of argument and dogmatism of belief of the type that Enlightenment thinking fundamentally rejected.

What some have termed the pacifism of Enlightenment thinking pertained empire as well. How much did this critique of the war system reflect the sharp resistance of colonized peoples? Recently, David Graeber and David Wengrow have argued in their much discussed book, *The Dawn of Everything: A New History of Humanity*, that Enlightenment critiques of society were deeply influenced by Indigenous thinkers rejecting the predations of empire, whose views constituted what Graeber and Wengrow call an "Indigenous critique."[23] I leave the assessment of this to intellectual historians and will take Diderot as my case in point, without prejudice as to the sources of his thinking.

Skepticism of empire was widespread among European thinkers in the late 18th century. We shall see that this contrasts dramatically with later perspectives in the 19th century. Adam Smith and Jeremy Bentham questioned empire on economic, political, and moral grounds. Edmund Burke is remembered for his attack on British policies in India. But Diderot's challenge was the most radical. Genocide and slavery were the inevitable result of Europeans' "rage of extending their dominions." Such were the depredations wrought by empire that they seemed to be the result of an "evil genius," in his words.

> Settlements have been formed and subverted; ruins have been heaped on ruins; countries that were well peopled have become deserted; ports that were full of buildings have been abandoned; vast tracts that had been ill cemented with blood have separated, and have brought to view the bones of murderers and tyrants confounded with each other. It seems as if from one region to another prosperity has been pursued by an evil genius that speaks our [European] several languages, and which diffuses the same disasters in all parts.

In his eyes, Europeans were imposing the same ruin and havoc on others that the barbarian invasions had imposed on them, except on a more global scale. "Savage Europeans!" he wrote,

> You doubted at first whether the inhabitants of the regions you had just discovered were not animals which you might slay without remorse

---

23 Graeber, David and Wengrow, David. *The Dawn of Everything: A New History of Humanity*. Farrar, Straus and Giroux, 2021, p. 5ff.

because they were black, and you were white. . ... In order to repeople one part of the globe that you have laid waste, you corrupt and depopulate another.[24]

The political upheavals of the late 18th century were typical in how they resulted from crises in war building. "All of Europe's great revolutions," Tilly writes, "and many of its lesser ones, began with the strains imposed by war."[25] The English Revolution began with Charles I's efforts to bypass Parliament in raising revenue for war on the Continent and in Scotland and Ireland. The American Revolution began with Britain's imposing taxes on the Colonies for the debts of the Seven Years War. The French Revolution was prompted by the monarchy's debts from the Seven Years War and the War of American Independence. Like the Thirty Years War, the Napoleonic Wars involved a military revolution whose antecedents were already in place. The American Revolution involved an army driven by civic ideals. The Napoleonic revolution in warfare mobilized unprecedented number of soldiers via nationalist passion. The *levee en masse*, proclaimed to defend France against its enemies foreign and domestic, called on all men capable of military service to fight and everyone else to contribute as they could to the war effort. Armies in the Hundred Years War seldom exceeded 12,000 soldiers, and British forces in the War of Spanish Succession were around 70,000 soldiers. Napoleon took over 600,000 soldiers into his invasion of Russia. With this, past limits on wars were cast aside. The focus of military tactics moved from cautious wars of position to ones of large-scale decisive battles whose aim was not just destroying the enemy army but also overturning the enemy state. Notably, Napoleon suffered his greatest defeat in Russia when he was unable to get the Russians to commit to a decisive battle in defense of Moscow. The emergence of this new, nationalized war foreshadowed the full emergence of that new entity, the *nation*. Bobbitt writes that for the first time we can speak of a war not just of kings but self-consciously of nations.

[2]/ Total War

[a]/ Nation-States and Free Trade Empires

Enlightenment critics of war attributed it to the machinations of detached rulers and ruling classes, such that once "the people" took over it would cease. What these critics did not appreciate was how deeply implicated the

24 Muthu, Sankar. *Enlightenment against Empire*. Princeton University Press, 2003, pp. 172–171.
25 Tilly, *Coercion, Capital, and European States: AD 990–1992, Op. cit.*, p. 186.

state was in war. Specifically, they did not foresee how the state could enlist democratization for war via the new ideology of nationalism and a new form of the state—the nation-state.

The political settlement following the Napoleonic Wars reinstated the dynastic state but alongside the nationalism which those wars had generated. The upshot was a century-long conflict unresolved until World Wars I and II over which principle would prevail: the older dynastic state or the new nation-state. But I speak of this as the era of the nation-state because as early as 1848 the fate of the dynastic state was clear, especially to its rulers who sought to contain the forces of nationalism.

At the heart of this new type of state was a new version of what I call the war contract.

Sociologist Andreas Wimmer puts it thus:

> Increasing state centralization and military mobilization led to a new contract between rulers and ruled: the exchange of political participation and public goods against taxation and the military support by the population at large. The idea of the nation as an extended family of political loyalty and shared identity provided the ideological framework that reflected and justified this new compact.[26]

The paradoxical relation of liberalism and militarism persisted, and in heightened form. Before, privileged groups received rights in return for their support in war. Now, rights were granted to more and more people as "subjects" became "citizens"; and the rights became more and more substantive—formal political rights, in the 19th century, were augmented by welfare rights, in the 20th century. All of which was the reward for "military support" in Wimmer's words, thus making the powers of war building all the more expansive.

And with this, the problem of mobilizing and managing the powers of destruction assumed new and heightened forms.

If the dynastic state faced challenges of motivating soldiers to fight, a challenge for the nation-state was *constraining* them from fighting for reasons of their own independent of the state's—or even contrary to the state's. The problem is framed by John Keegan. The challenge, he writes, faced by rulers immediately after the Napoleonic Wars was "How might one have the forms of warfare practiced by the armies of the French republican Napoleon without the politics of revolution?"[27] This was brought

---

26 Wimmer, Andreas. *Waves of War: Nationalism, State Formation, and Ethnic Exclusion in the Modern World.* Cambridge University Press, 2013, p. 18.
27 Keegan, *Op. cit.*, p. 17.

dramatically to the fore by the 1871 Paris Commune, an event that looms large in the history of socialism and is equally significant for modern war. Bismarck had launched a war of national consolidation against France, one of several such wars that created modern Germany. But the people of Paris refused to accept their government's surrender so took up arms in their own defense—and in so doing sought to cast aside remnants of the old regime that had lingered from their past military prestige. As Karl Marx noted in his account of the Commune, a main demand of the Communards was the abolition of standing armies. The Paris Commune was ultimately crushed but its legacy endured in the specter of how nationalized war could ignite popular revolution. (Charles DeGaulle notes in his World War II memoirs that one reason the French military so quickly capitulated to Germany was fear that a protracted struggle would again occasion something like the Paris Commune.)

Framing it thus suggests posing the problem of mobilizing and managing the powers of war as one of the relations *between* the state and the nation. I have spoken of the "nation-state" in keeping with common parlance and will continue to do so. But the relation between them is problematic in ways that render war itself problematic; indeed, their tension and often conflict has been a defining feature of the 19th and 20th centuries, if not still. Someone who addressed this was Clausewitz. Writing after the Napoleonic Wars, he was the first to theorize the *novelty* of nationalized war. What do we mean when we talk of the "nation-state" and its "aims" in war? Clausewitz held that war was actually fought by what he called a "trinity" of factors: the government, the military, and the people, which he also termed the nation. His famous claim that war was political in essence was partly to stress that at given time the parties to this trinity were as much in conflict with each other as an agreement, to the point that one could question if they constituted a single agent. As noted in the previous chapter, Clausewitz went so far as to suggest that its status as a single agent was, in fact, a myth.

Rousseau posed the question of the artificiality of the state. The question now raised concerned the status of the "nation."

Wimmer notes that nationalism "has provided the ideological motivation for the increasing number of wars fought in the modern era."[28] Indeed, if one had to settle on a single factor to explain why over 3,000 people a day died from wars in the 20th century, a good choice would be nationalism. It has been the motivating ideology of interstate wars. And it has driven the two other major forms of war in our time: war of state *consolidation*, where an existing state imposes a single national identity on an

---

28 Wimmer, *Op. cit.*, p. 14.

otherwise heterogeneous population (leading to repression, civil wars, and in the extreme genocide—such as Turkey's genocide of the Armenians); and wars of state *formation*, where a state is grafted onto a previously politically amorphous region of jumbled polities (post-World War I Middle East and post-World War II Africa are cases in point). And the lines between these conflicts have been blurry. Wars between states have involved wars within states, so that in the 20th century you were just as likely to be killed by your own government as by an "enemy" one.

What are we to say, then, about the fact that to this day no one can agree on what a "nation" is—beyond being something you fight and die for?

The nationalist wars of the 20th century have been rightly compared to the religious wars of the 17th century. But people can at least reach some agreement on what a "religion" is. Our leading scholar of nationalism, Benedict Anderson, acknowledges, "Nation, nationality, nationalism—all have proved notoriously difficult to define, let alone to analyze."[29] Eric Hobsbawm states bluntly that while attempts have been made "to establish objective criteria for nationhood . . . [a]ll such objective definitions have failed."[30] As the 19th-century-British political theorist Walter Bagehot lamented, "But what are nations? What are these groups which are so familiar to us, and yet, if we stop to think, so strange?"[31]

Scholars of nationalism distinguish *civic* nationalism, for which the nation is a political entity constituted by shared values; *cultural* nationalism, for which the nation is a cultural entity typically identified with a common language as the bearer of a shared past; and *organic* nationalism, where the nation is conceived as a kind of collective organism with individuals being appendages of them. (This is sometimes termed *ethnic* nationalism.) It is hard to identify anything these notions have in common beyond their relation to war. Hence, Sir Michael Howard has suggested that the nation simply cannot be conceived independent of war. In his words, "It has been characterized almost everywhere by some degree of militarism." Indeed, "It is in fact very difficult to create national self-consciousness without a war."[32] As Anderson has noted, the nation is above all else something that

---

29 Anderson, Benedict. *Imagined Communities: Reflections on the Origin and Spread of Nationalism.* Verso, 2006, p. 3.

30 They tend to be flatly circular. Hobsbawm endorses Masaryk's definition of nationalism as "any outlook that treats the nation as the highest political value." But this just defines "nationalism" in terms of "nation." Hobsbawm, Eric. *Nations and Nationalism since 1780.* Cambridge University Press, 2012, pp. 10 and 12.

31 Bagehot, Walter. *Physics and Politics: Or Thoughts on the Application of the Principles of Natural Selection and Inheritance to Political Society.* Legare Street Press, 2021, p. 83.

32 Howard, Michael. "War and the Nation–State." *Daedalus*, 108(4), 1979, pp. 101–110.

you die for. Other collective identities may impact you more, like your university, but no one "gives their life" for their university—they die for their nation. And the nation serves as a vehicle of mobilization both by inclusion and exclusion. The mobilization of the populace by transforming them from subjects to citizens also involves the oppression if not elimination of those seen standing in the way of the national effort, or those who are usefully stigmatized to energize that effort.

The rise of nationalism also motivated a dramatic change in attitudes toward imperialism.[33] Critiques of imperialism were prominent in the late 18th century, as we have seen. But just a half-century later, no prominent political thinkers in Europe questioned the justice of European empires. Support was especially notable among liberal-minded thinkers like Alexis de Tocqueville, James Mill, and John Stuart Mill, who regarded "despotism" as a "legitimate mode of government in dealing with barbarians." The changed attitudes laid the groundwork for the massive imperial conflicts of the 20th century.

With the Industrial Revolution, empire was no longer about luxury goods like sugar and tea but now became an integral part of the European economy as its raw materials were needed to feed the machines of Europe and its inhabitants were needed as markets for European goods. A case in point was the need of the British textile industry for cotton, but a better example was India which became a model for all European colonies in the 19th century. It was a source of raw materials as well as a market for British textiles after India's own textile producers were repressed. Imperialism now reached its full potential in what Diderot had condemned as destroying local cultures and systems of government. And racism was becoming part of its official ideology, as local ethnic and cultural differences were exploited to keep the colonized divided and the colonizers increasingly segregated themselves from the locals as a mark of their white superiority.

Previously, empire, like war generally, had been detached from the ordinary person's concern. Now it became a central part of nationalist ideology. Religious proselytizing receded as a motive for empire as nationalist pride became entwined with empire building, including the conviction that nations were bringing "progress" to the realms they controlled, a central feature of liberal imperialist ideologies like John Stuart Mill's. It also played a significant domestic role. Nineteenth-century industrialism brought large-scale class conflict to Europe for the first time. One response to such discord was an ideology of national unity grounded in imperialist

---

33 Pitts, Jennifer. *A Turn to Empire: The Rise of Imperial Liberalism in Britain and France.* Princeton University Press, 2005.

endeavors as ones in which the populace as a whole was implicated, and in which it could find fulfillment.

Finally, empire became even more integral to war making and war building. Britain exemplified this. Though the 19th century's dominant military power, Britain's domestic military expenditures were quite low—only 2 to 3% of GNP after 1815. But this was because war building was largely externalized. Half of all soldiers under British control were in India, ready to be dispatched elsewhere—for example, to China in 1860 and again in 1900–1901; to Sudan in 1885 and 1896; and to Egypt in 1882. Remarkably, two-thirds of the cost of all British soldiers were funded by Indian taxpayers! (No wonder that Britain's military priority through World War II was maintaining sea-lane access to India.) All of which allowed the British to pride themselves on their lack of militarization compared to the Continent while being the dominant power globally.

[b]/ The Deluge

The great American sociologist W.E.B. Du Bois famously maintained that the origins of World War I lay in Africa and the so-called scramble for Africa that obsessed late 19th-century Europe. He related it to the fate of other empires:

> Nearly every human empire that has arisen in the world, material and spiritual, has found some of its greatest crises on this continent of Africa, from Greece to Great Britain. As Mommsen says, "It was through Africa that Christianity became the religion of the world." In Africa the last flood of Germanic invasions spent itself within hearing of the last gasp of Byzantium, and it was again through Africa that Islam came to play its great role of conqueror and civilizer.

Yet the centrality of empire remained foreign to Western consciousness:

> The methods by which this continent has been stolen have been contemptible and dishonest beyond expression. Lying treaties, rivers of rum, murder, assassination, mutilation, rape, and torture have marked the progress of Englishman, German, Frenchman, and Belgian on the dark continent. The only way in which the world has been able to endure the horrible tale is by deliberately stopping its ears and changing the subject of conversation while the deviltry went on.[34]

---

34 Du Bois, W.E.B., "The African Roots of War." *The Atlantic*, May 1915. www.theatlantic.com/magazine/archive/1915/05/the-african-roots-of-war/528897/.

The realities of World War I were lost in a haze of ideological abstractions. Hew Strachan puts it thus: "Big ideas, however rhetorical, shaped the war's purpose more immediately and completely than did more definable objectives."[35] No one had bothered to ponder the "larger meaning" of 18th- or 19th-century conflicts like the War of Jenkins Ear or the Crimean War. Now both sides sought the Cosmic Significance of all those deaths in the mud of Passchendaele or all those starved by Britain's economic boycotts. Wars of empire had traditionally been portrayed in such terms, so this was another instance of empire's logic returning to Europe. Religion was enlisted to portray the stakes in ways not seen since the 17th century's religious wars. One Protestant pastor proclaimed the war a "new Whitsun" [Pentecost—CR], "the coming of the Holy Ghost 'like a mighty, rushing wind.'" Cornell University's president announced that the war was a struggle between "the kingdom of heaven and the kingdom of Hunland."[36] The day Germany declared war, German historian Friedrich Meinecke proclaimed that it marked a "new historical epoch for the world." Even such a lumbering thinker as Edmund Husserl found metaphysical meaning in the conflict. Not to be outdone, French thinkers like Henri Bergson defined the conflict as between German "barbarianism" ("her brutality, her appetites, and her vices") and French "civilization."[37]

On one level, World War I was a rather adolescent conflict over "prestige." This was not without its domestic importance, as "prestige" and similar concerns played a role in mobilizing nationalist sentiments to quell domestic discord. On another level, "prestige," etc. were so closely identified with the possessions of an empire that colonial conflicts achieved an importance they had not previously had. Ironically, part of the "prestige" associated with empire was the notion that Europeans were doing a valued service in bringing their blessings to others; there was a competition in altruism, as it were. But alongside this was a contrary, if equally racist assumption. This was the late 19th-century Social Darwinist notion that nations were engaged in a "survival of the fittest" from what some would emerge and others not, and that the "nations" were racially constituted entities that would demonstrate their superiority by their survival. Finally, the contending alliances in the conflict, especially that of Britain, France,

---

35 Strachan, *Op. cit.*, p. 1114.
36 Strachan, *Op. cit.*, p. 1116.
37 See de Warren, Nicolas and Vongehr, Thomas, editors. *Philosophers at the Front. Phenomenology and the First World War.* Leuven University Press, 2018; de Warren, Nicolas. "The First World War, Philosophy, and Europe." *Tijdschrift voor Filosofie*, 2014, pp. 715–737.

and Russia, would not have made any sense absent their former consolidation in colonial conflicts.

But what were the "colonies"?

It bears stressing that, for the Central Powers (Germany and Austria-Hungary), parts of Eastern Europe were regarded as possessing colonial status. This is why the conflict in Serbia that sparked the war was regarded by both Russia and Austria-Hungary as speaking to their spheres of influence as major powers. More important in the long run was the German attitude toward the lands to its east. Every imperialist power saw its colonies as sources of raw materials as well as places where its "excess population" could be relocated. Long before Hitler's talk of *lebensraum* (living space), that notion had been employed to denote the so-called primitive so-called lesser developed lands to Germany's east that Germany could claim for the same reasons that others claimed lands in Africa. It was this perception of parts of Europe as themselves ripe for colonization that made the world wars the third-hyper war involving the attempt to impose an empire on all of Europe.

It is in popular culture that we often find the most perceptive critiques of 20th-century warfare. I shall comment later on the invisibility of the threat posed by the ongoing manufacturing of nuclear weapons post-World War II. That threat was addressed in the science fiction movies of that era where nuclear testing was portrayed as unleashing on unsuspecting communities a parade of giant ants, giant grasshoppers, giant scorpions, etc., or impacting humans so that they grew to monstrous sizes or shrank to sub-molecular size. Some of the most penetrating explorations of the madness of World War I were in horror films, a genre to which the conflict gave rise. The horror film's most prominent director, James Whale, fought on the Western Front and was a prisoner of war. He passed his time cultivating the theatrical skills that he would later employ in films like "Frankenstein" and "The Bride of Frankenstein." Whale's "The Invisible Man," based on the H.G. Wells novel, was adapted by R.C. Sheriff, who had previously authored the antiwar play "Journey's End." Its protagonist's face is entirely bandaged (like victims of World War I facial injuries); he is otherwise invisible (like World War I's "forgotten men"), and his insanity ultimately drives him to murder. The horror genre originated with the 1920 German film, *The Cabinet of Dr. Caligari.* Regarded as the quintessential work of German expressionist cinema, it was a major influence on Whale and others in the horror genre. It was written by two pacifists, Hans Janowitz and Carl Mayer, who translated the insanity they experienced in the war into a parable of authority and obedience involving an insane hypnotist (i.e., the warring governments) who employs a somnambulist/sleepwalker (i.e., the obedient populations) to commit senseless murder.

World War II was an even more straightforward imperialist conflict, but now driven by what has been called "catastrophic nationalism."[38] Empire remained the mark of a great power, as the dénouement of World War I had consolidated and extended the empires of the victors. As much as anything, Germany resented the loss of its colonies and its great power status in the Versailles settlement. From its perspective, all the talk of national self-determination had only led to the creation of new states in Europe that had divided the German peoples from one another. The other factor was the Great Depression, itself largely the product of forces resulting from World War I. The protectionist policies that all countries adopted in response to it convinced Germany and Japan that the only way they could secure the raw materials they needed for their survival, much less for their flourishing, was through the kind of imperial policies they identified with Britain. Hitler himself never thought that war with Britain was necessary. He did not see why he could not have Russia just as Britain had India. But Britain had always regarded a balance of power in Europe as a prerequisite of its empire's security which is why, when Germany invaded Poland—a country of insignificance to Britain and its people—Churchill could maintain that the war was about the defense of the British Empire.

[c]/ Post-World War II and Martial Liberalism

The post-World War II era was the high point of the warfare–welfare state and the militarization of liberal democracy as its ideology.[39] I have termed this ideology *Martial Liberalism*, at whose heart is a generalization of the citizen soldier ideal.[40] On the one hand, all citizens are soldiers, meaning states can make exorbitant claims on them in the name of national defense; on the other hand, all soldiers are citizens, meaning they are entitled to expansive political and social rights. Western countries witnessed the expansion of political and social rights and the increasing "civilianization" of the state. The American Civil Rights Movement was prompted partly by African American participation in the war as well as America's Cold War imperative of establishing itself as a beacon of freedom. In Britain, the shift to social spending was dramatic. In 1898, the biggest budget item had been the 36% devoted to the military with social services being only

---

38 See Geyer, Michael. "'There Is a Land Where Everything Is Pure: Its Name Is Land of Death': Some Observations on Catastrophic Nationalism." *Sacrifice and National Belonging in Twentieth-Century Germany.* Edited by Greg Eghigian and Matthew Berg, Texas A&M University Press, 2002, pp. 120–141. Cited in Overy, *Op. cit.*, p. 35.

39 For a general discussion of the United States, see Sherry, Michael. *In the Shadow of War: The United States since the 1930s.* Yale University Press, 1995.

40 See Ryan, Cheyney. *Chickenhawk Syndrome, Op. cit.*

10%. By the end of the 20th century, social security, health, and education constituted over 50% of the budget and the military only 7%. But historian David Edgerton would still title his influential study of Britain in the Cold War *The Warfare State*.[41] In America, defense spending tripled with the onset of the Cold War, and the war-induced paranoia of McCarthyism engendered massive peacetime infringements on liberties.

Post-World War II global arrangements were also understood on the model of the warfare–welfare state. During the war, President Franklin D Roosevelt envisioned an American global posture modeled on a New Deal for the world, of massive government spending for both security and progress. Events soon transformed Roosevelt's "one-world" vision into Truman's project of American hegemony in which America's allies received material benefits of various sorts in return for participation in/support for America's global aspirations—starting with containment of the Soviet Union. America consolidated its position as the dominant power with an empire that was even more effective than that of the British in insulating the American people from their country's military endeavors. What most distinguished it from its British forerunner was the paramount role of overseas bases. The half-dozen naval stations America possessed in 1900 grew to 300 overseas bases by the mid-1950s and to around 800 in 70 countries by the end of the Cold War. It became, in Daniel Immerwahr's words, a "pointillist empire" involving a "new pattern of global power, based less on claiming large swaths of land and more on controlling small points."[42]

Industrial warfare was consummated in nuclear weapons, the most significant development in war building since gunpowder. They exemplified all that had ever been worrisome about war building. They were the *reductio ad absurdum* of where war-oriented society had been headed all along in several respects:

- They exemplified war-building's cost. The Manhattan Project was the largest public works project in American history, costing six times more than all the transcontinental railroads combined. It was conducted at 30

41 Edgerton, David. *Warfare State: Britain, 1920–1970*. Cambridge University Press, 2006.
42 See Immerwahr, Daniel. *How to Hide an Empire: A History of the Greater United States*. Picador, 2020, p. 2016. See also Vine, David. *Base Nation: How U.S. Military Bases Abroad Harm America and the World*. Metropolitan Books, 2015. These discussions build on Chalmers Johnson's trilogy: *Blowback: The Cost and Consequences of American Empire* (2000), *The Sorrows of Empire: Militarism, Secrecy, and the End of the Republic* (2004), and *Nemesis: The Last Days of the American Republic* (2006)—all published by New York: Metropolitan Books. See also Vine, David. *The United States of War: A Global History of America's Endless Conflicts, From Columbus to the Islamic State*. University of California Press, 2020.

sites nationally employing over 130,000 workers, at a cost the equivalent of $30 billion in today's dollars.

- They exemplified its invisibility. The cost of the Manhattan Project was concealed from Congress. Not even the American Vice President Harry Truman was told about its existence. Most involved did not know the purposes of the project. And the hiddenness persisted after the war. Few knew of the Hanford Nuclear Reservation to process uranium into plutonium, though at one point it consumed almost 10% of the nation's electricity.
- Finally, they exemplified authoritarian control over the war power to an absurd degree. Gary Wills and others have argued that this has enabled centralized power on every level.[43]

In the most penetrating analysis of their political significance, *Thermonuclear Monarchy: Choosing Between Democracy and Doom,* Elaine Scarry has argued that nuclear weapons are incompatible with the very notion of governance. Sober analysis reveals that they simply "cannot be kept securely within a legitimate command structure"; they are essentially "wholly outside human control."[44] She cites an article claiming "that robots may eventually have so much intelligence and power built into them that they will be able to disable their human makers before those human makers can disable them."[45] They are the culmination of the inexorability of the war system present from the start.

Like gunpowder, the impact of nuclear weapons on war and the societies that wage war has been profound. As noted in Chapter 1, the Cold War witnessed conflict move away from direct encounters between the superpowers to the so-called non-Western world whose conflicts were proxy wars between the superpowers. Indeed, the Cold War was only "cold" for the major powers. But a variety of factors were eroding nationalism and with it nationalized war. Some of them resulted from technological changes. Whereas firearms had led to the emergence of mass armies and eventually the militarization of the entire populace, nuclear weapons led in the opposite direction. The link between citizenship and soldiering was

---

43 Wills, Gary. *Bomb Power: The Modern Presidency and the National Security State.* Penguin, 2011; Hoffman, David. *The Dead Hand: The Untold Story of the Cold War Arms Race and Its Dangerous Legacy.* 1st ed. Anchor, 2010, pp. 143–155; Rhodes, Richard. *Arsenals of Folly: The Making of the Nuclear Arms Race.* Deckle Edge, 2007.

44 Scarry, Elaine. *Thermonuclear Monarchy: Choosing Between Democracy and Doom.* Summary ed. W. W. Norton & Company, 2016, p. 259.

45 Scarry, *Thermonuclear Monarchy: Choosing Between Democracy and Doom,* p. 259, referencing Joy, Bill. "Why the Future Doesn't Need Us: Our Most Powerful 21st Century Technologies—Robotics, Genetic Engineering, and Nanotech—Are Threatening to Make Humans an Endangered Species." *Wired,* 8(4), 2000. Online at www.wired.com/2000/04/joy-2/.

severed, as large armies were no longer needed. Indeed, the link between war and shared sacrifice so crucial to nationalism no longer made sense except insofar as a nuclear exchange meant everyone would passively die. This was accompanied by large-scale disillusionment as the realities of 20th century war sunk in. The sheer extent of killing and destruction in World War II conjoined with the prospect of nuclear holocaust and the various forms of subterfuge in which Cold War governments engaged, whose immorality they sought to keep hidden: all contributed to what I have termed the Great Disenchantment.

The militarization of liberal democratic ideology, so complete in the immediate post-World War II world, also unraveled. This mainly reflected the erosion of nationalism. Liberalism had reconciled its ideals with acceptance of militarist arrangements like conscription by accepting the claims of the nation state as necessary to the defense of those ideals. But once the ideal of the nation started eroding, liberals began critiquing those arrangements starting with conscription in the name of individual rights. The same held true for the claims of democracy. Democratic ideals could live with the intrinsically undemocratic conduct of war making and war building as long as the exigencies of the nation were taken to demand it. Once those exigencies were questioned, democratic doubts about war followed.

It was not surprising that a theme in post-World War II political philosophy became the critique of nationalism. It was central to one of the most popular works of the era, Karl Popper's *The Open Society and Its Enemies*.[46] He articulated a view shared by other liberal-minded thinkers that Nazi-ism was an expression of ultra-nationalism. But instead of seeing nationalism as what it was, an expression of modernity's intimate relation to war, his critique lapsed into the familiar categorizations of European imperialism, if not racism. Nationalism was a reversion to the "primitive," the "tribal"; it was as if Hitler had emerged from the jungles of "darkest Africa" instead of the most enlightened country in Europe. The upshot was diverting attention from the true source of the problem: war. This continuing inattention to war that has permitted subsequent liberal thinkers to reconcile their ideals with it.

[IV]/ After the Cold War: Alienated War, "New Wars," and Terrorism

So if the immediate post-World War II era saw the height of the war contract in Martial Liberalism, the subsequent years saw its unraveling leading to the Great Defection.

---

46 Popper, Karl. *The Open Society and Its Enemies*. Princeton University Press, 2020.

Things took longer to unravel in the United States than in postwar Europe, but when they did, they did so dramatically. The Vietnam War demonstrated again the democratizing impact of nationalized war as the vote was extended to 18-year-olds. But it also demonstrated its radicalizing impact as the war engendered America's greatest political crisis in a century. Civilian disenchantment was evidenced in the flat-out refusal to serve. 570,000 young men were classified as draft offenders, and over 200,000 were formally accused of draft violations (adjusted for population growth, this would be equivalent to 900,000 and over 340,000 today). Even more dramatic was the turmoil in the military. Department of Defense figures reported over 500,000 desertions between 1966 and 1973 (compared with 13,000 during the Korean War).[47] The Tet Offensive has been identified as when combat avoidance turned to outright mutiny. *Armed Forces Journal* stated in 1971 that

> Our army that now remains in Vietnam is in a state approaching collapse, with individual units avoiding or having refused combat, murdering their officers and noncommissioned officers, drug-ridden, and dispirited where not near-mutinous. [C]onditions [exist] among American forces in Vietnam that have only been exceeded in this century by . . . the collapse of the Tsarist armies in 1916 and 1917.[48]

President Nixon's Chief of Staff H.R. Haldeman informed him, "If troops are going to mutiny, you can't pursue an aggressive policy."[49]

Writing in the immediate shadow of the Vietnam War and the waning of nationalized war/martial liberalism, Sir Michael Howard asked: "What next?"

His answer was a mordant return to the 18th century. Post-nationalized war would be denationalized to become what it once was, an affair of states rather than peoples: "The State apparat is likely to become isolated from the rest of the body politic, a severed head conducting its intercourse with other severed heads according to its own laws."[50] The alienation would be several fold. The citizenry would be less willing to participate in war or

---

47 Accounts of this era are in Cortright, David. *Soldiers in Revolt: The American Military Today.* Doubleday, 1975; Wells, Tom. *The War Within: America's Battle Over Vietnam.* Henry Holt, 1994; Boyle, Richard. *GI Revolts: The Breakdown of the U.S. Army in Vietnam.* United Front Press, 1972; Appy, Christian G. *Working-Class War: American Combat Soldiers and Vietnam.* University of North Carolina Press, 1993.

48 Heinl, Jr., Colonel Robert D. "The Collapse of the Armed Forces." *Armed Forces Journal,* 7 June 1971, reprinted in Gettleman Marvin, et al. *Vietnam and America: A Documented History.* Grove Press, 1995, p. 327.

49 Cited in Wells, *Op. cit.*, p. 475.

50 Howard, Michael, "War and Nation State." *Op. cit.*

contribute in any significant way. The actual fighting of wars would be increasingly privatized via the rise of corporate warriors, today's equivalent to the professionals/mercenaries of the 18th century. Also, wars would be paid for by borrowing, a return to the financial novelties of the 18th century. But some of the revulsion at total wars would remain. Citizens would be less tolerant of immoral state practices resulting from blind patriotism. If states were to wage war, they would have to convince the populace that they were doing so morally, or in the words of Samuel Moyn "humanely."

At the same time, political leaders would become less patient with policies impeding their actions since they would no longer be required to ensure popular participation. They would seek more ways to insulate what they did from popular assessment.[51] Sir Michael was writing in the shadow of the Watergate scandal, and a few years before the Iran-Contra scandal. Both were major American constitutional crises of the 20th century involving abuses of executive war powers. Later, both the United States and the United Kingdom would witness abuses of executive authority in the decision to invade Iraq.

Ironically, then, democracy's triumph over totalitarian forces meant that war would become increasingly *un*democratic.

What Sir Michael saw was a reversion to the past. But post-Cold War discussions would be characterized by themes of newness, indeed of "revolution" in how wars would now be fought. The end of the Cold War saw much discussion of what came to be called the "Revolution in Military Affairs" or RMA. The basic idea was that dramatic technological advances, such as in computational power, the size of computer components, high-tech intelligence gathering techniques and an array of other developments would dramatically change the nature of battle. Of special importance was the perceived quantum leap in the reach, efficacy, and lethality of airpower, conjoined with information capabilities that would all but eliminate the "fog of war" that had characterized armed conflict since ancient times. In the American military, these "force multipliers" were packaged with a constant multiplication of new terms/initials: Airborne Warning and Control System (AWACS), Joint Surveillance Target Attack Radar System (JSTARS), Joint Tactical Information Distribution System (JTIDS), and so on. The new techno-vision fit with the downsizing of military forces in portraying the armies of the future not as the sluggish massive forces of the past but as smaller, faster, more "agile" units.[52]

51 Savage, Charlie. "U.S. Discloses Decades of Justice Dept. Memos on Presidential War Power." *New York Times*, September 16, 2022. www.nytimes.com/2022/09/16/us/politics/war-powers-justice-dept-president.html.
52 Chapman, Gary. "An Introduction to the Revolution in Military Affairs." *Proc. XV Amaldi Conference on Problems in Global Security*, Helsinki, Finland, September 2003.

It turned out that this "Revolution in Military Affairs" worked well as long as the enemy fought traditional wars. Such was the case in the first Gulf War when Saddam Hussein and Iraqi forces chose to fight the United States in what scholars describe as the last World War II battle. This did not deter proponents of RMA from proclaiming that the first Gulf War gave us a vision of the future. But this vision was rudely shattered by the next Iraq conflict that confirmed, as the Afghanistan war would further confirm, that all this technological wizardry was useless in the long run. As in Vietnam, where American leaders had once proclaimed that their technological mastery would eventually prevail over the cruder methods of the adversary, the lesson that endured from Clausewitz was about the primacy of politics in war.

Was the post-9/11 "War on Terror" a "new" type of war? I noted in Chapter 1 that received thinking was ambivalent. On the one hand, pundits likened America's predicament to that of Rome against the barbarians or European settlers against Native Americans. But the parallels with past colonial wars went even further to include how to understand the threat of terrorism.

The emergence of "terrorists" as a distinct category is part of the story of the consolidation of the state.[53] Generally, they have been groups that were neither appropriated into nor annihilated by the process of state formation but relegated to the margins where they served as both an extension of and a challenge to the state's monopoly of force. A case in point is piracy. So-called buccaneers emerged in the 17th century as "Brethren of the Coast" (one of their designations), independent sailors/ships, many of them Europeans transplanted to the Caribbean. They were initially enlisted as independent contractors by England and France to harass Spain, giving them legal status as so-called "privateers." In this role, they were heroes, lauded in popular culture and even knighted. But the monopolization of the state over all warfare eventually meant that they were redefined as pirates and branded as criminals whose very existence posed a challenge to the state system. Scholarship today takes special interest in the distinct political societies that pirates formed, a kind of martial counterculture that was Spartan in its egalitarianism.[54]

---

53 Bobbitt, Phillip. *Terror and Consent*. Anchor Books, 2008.
54 Linebaugh, P. and Rediker, M. *The Many-Headed Hydra: Sailors, Slaves, Commoners, and the Hidden History of the Revolutionary Atlantic*. Beacon Press, 2013; Rediker, M. *Between the Devil and the Deep Blue Sea: Merchant Seamen, Pirates and the Anglo-American Maritime World, 1700–1750*. Cambridge University Press, 1989; Rediker, M. *Outlaws of the Atlantic: Sailors, Pirates, and Motley Crews in the Age of Sail*. Beacon Press, 2015.

One of their more significant roles was at the margins of "civilized"/"savage" warfare. The European claim to fight according to the laws of warfare was part of their general claim to superiority over non-Europeans. But this often meant enlisting others to do their dirty work for them; hence, the role of "irregular" fighters has been a persistent feature of modern warfare. One example is the enlistment of Indigenous tribes by European powers in America in their conflict with other Indigenous tribes. Britain and France used the Iroquois and the Algonquin in what has been called a first instance of state-sponsored terror. Every conflict between the United States government and Native American tribes involved the enlistment of native peoples, a practice that continued in the American occupation of the Philippines and the guerrilla war that followed. It had the benefit of exacerbating divisions between native tribes in ways that facilitated their ultimate subjugation. Such irregular forces were also employed in Europeans conflicts with each other. Think of Cossacks, Highlanders, and the Hussars, who all played such roles. Magyar light cavalrymen were recruited from Hungary and Christian refugees (loosely termed as "Albanians") from the Ottoman Balkans. (The latter are figures in Mozart's "Cozy Fan Tuttle" where they exemplify the mystery around such wild figures.) It takes little imagination to see here anticipations of what America confronted in its "War on Terror." In the waning decade of the Cold War, the practice of sponsoring native irregulars to harass an opponent on the frontier found a clear analogue in American sponsorship of Islamic groups fighting the Soviet Union in Afghanistan. Al Qaeda, we might say, began as the equivalent of privateers then later transformed into the equivalent of pirates, also creating their own independent society from which to attack the so-called civilized world.

Finally, it seems to me that what scholars have debated as "New Wars" are as much as anything an instance of what I have portrayed as the "chickens coming home to roost."

Mary Kaldor has identified two logics in today's wars.[55] One is "New Wars" proper, conflicts occurring in the context of weak states, financed by outside sources or criminal activities, and organized around issues of so-called "identity." Another word for this is "degenerative" war as it has none of the state-building consequences of war plus its focus on identity means that the military target is often a whole community. There are some new elements here. The organizations involved increasingly resemble Visa or MasterCard organizations, their activities blurring all distinction between war and crime, plus there is the possibility of them acquiring means of mass

---

55 Kaldor, Mary. *New and Old Wars: Organized Violence in a Global Era*. 3rd ed., Stanford University Press, 2012.

destruction so that their powers to inflict death are exponentially magnified. But it is hard to see how this is really much different than the logic of hostilities as it used to occur on the margins of the European world, where war has had the same degenerative impact—except now, the threats are returning to the Euro-American world.

Along with this, Kaldor writes, are the media-friendly high-technology "spectacle wars" that the developed countries have waged. Since the end of the Cold War, state military forces have been increasingly privatized, reversing the process of socialization identified with the nation-state. This is part of a larger movement of the commodification of killing generally. (A Marxist might see this as the ultimate triumph of capitalism and its penchant to commodify everything.) Such wars may aim to provide a semblance of the solidarity at home that real wars used to provide, but they inevitably fail. A fundamental reason is how they have severed the traditional link between nationalism and sacrifice so essential to the religious-type aura surrounding the claims of the nation. This may be the ultimate sign of nationalized war's demise. Hence, we have the strange phenomenon in the United States and Western European countries of political constituencies proclaiming the priority of nationhood, of being "American," etc., without its being hinged to any intelligible shared purpose other than excluding/denigrating others. In the United States, they champion a President who brags about avoiding military service and cheating on his taxes. The question is what happens to democracy since the era of nationalized war so intimately linked democracy and its claims to participation in the mission of the nation-state.

What all these forms of war continue to share is a reason to continue fighting.

[V]/ Disarming Peace

Why has contemporary politics become so unhinged? Why is the fundamental legitimacy of so many institutions being questioned, and why is consensus around so many basic issues vanishing?

My account of the war system provides an answer. War has been the cement that has held states together. States, empires, and nations have not just engaged in war, they have been *constituted* by war. Take away war, and notions like national community become detached from their traditional moorings. Thus in the United States, "patriotism" becomes compatible with shirking military service, avoiding taxes, undermining the institutions of justice, and ignoring the civic responsibilities that used to be what made the country worth fighting for. "Putting America first" means putting *oneself* first in ways that are devoid of any larger social purpose. The only shared commitment is keeping "non-Americans" out.

This simplifies things of course. Philip Bobbitt has emphasized the inability of today's state to realize its most fundamental claims to provide protection. Sir Michael Howard has put it in terms of the state's relation to the nation. Insofar as the state claims to "protect" the nation, it no longer needs the nation's active engagement to do so; at most, it needs its acquiescence. The relation between them is increasingly attenuated. War continues to be what the state is about, but war's existence for society writ large is a ghostly one, constituted by constantly recycling old images from the past. Every war is World War II *redux*: every conflict is an "existential threat" (though few can say even where the conflict is); every "adversary" must be stopped in its tracks as otherwise everything the nation "stands for" will be lost (though few agree on what the nation "stands for"). War is not just wrapped in nostalgia. Its promises a return to The Good Old Days, but without any of the sacrifices that past conflicts involved.

How do we make the dangers here present to people? I shall turn to this in the final chapter. The detachment of the state from the nation means that its war making capacities and proclivities proceed independently of public awareness and control. Technological developments are focused on increasingly insulating the public from the realities of war, as in the removal of the human element entirely from war fighting via drones and other forms of automatic weapons. Publics are no longer the agents of war but they much remain their potential victims. But there is also opportunity in the detachment of the nation from the state. Whatever benefits societies received from the war contract have vanished. The social and political rights citizens received for their involvement in the state's wars are a thing of the past. The "national pride" that citizens once found in the "glories" of their empires is gone, not because empires have gone but because their existence is too invisible or too ephemeral.

The challenge of reconceiving the nature of political community is also one of reconceiving peace. How much have our conceptions of peace been implicated in war, or have even enabled war?

Eighteenth-century critics of the war system raised the question of whether prevailing conceptions of peace were really deep enough. Kant maintained that a peace treaty that left mistrust and the sources of hostility in place was no peace treaty at all, it was just a temporary suspension of the fighting. By this logic, there has never been any real peace making. Writing in the same tradition, William James wrote at the start of the 20th century:

England and we, our army and navy authorities repeat without ceasing, arm solely for "peace," Germany and Japan it is who are bent on loot and glory. "Peace" in military mouths to-day is a synonym for "war expected." The word has become a pure provocative, and no government wishing peace sincerely should allow it ever to be printed in a newspaper. Every up-to-date dictionary should say that "peace" and "war" mean the same thing, now *in posse*, now *in actu*. It may even reasonably be said that the intensely sharp competitive preparation for war by the nations is the real war, permanent, unceasing; and that the battles are only a sort of public verification of the mastery gained during the "peace" interval.[56]

The official notion of "peace", he claims, is itself a "provocation" insofar as it names the intervals of war building in between the periodic bursts of warmaking. Indeed, those intervals are the "real war" because their permanent and unceasing character renders them the most impactful. The only solution is to create a peace that is truly disarmed.

Finally, consider this remark from the late 19th century, as Europe—believing it was at peace—frantically engaged in the project of war building that led to the disastrous 20th century.

The so-called armed peace that prevails at present in all countries is a sign of a bellicose disposition, of a disposition that trusts neither itself nor its neighbour, and, partly from hate, partly from fear, refuses to lay down its weapons. Better to perish than to hate and fear, and twice as far better to perish than to make oneself hated and feared—this must some day become the supreme maxim of every political community!— Our liberal representatives of the people, as is well known, have not the time for reflection on the nature of humanity, or else they would know that they are working in vain when they work for "a gradual diminution of the military burdens." On the contrary, when the distress of these burdens is greatest, the sort of God who alone can help here will be nearest. The tree of military glory can only be destroyed at one swoop, with one stroke of lightning. But, as you know, lightning comes from the cloud and from above.[57]

The author of this remark is, improbably enough, Frederick Nietzsche in *Human, All-Too-Human*. Nietzsche was certainly no pacifist, but perhaps anyone deeply responsive to the temper of the times could have sensed what he did around him. Armed peace was a peace of mistrust, hate, and

---

56 James, William. *The Moral Equivalent of War*. Obscure Press, 2013, p. 3.
57 Nietzsche, Friedrich. *Human All-Too-Human: A Book for Free Spirits, Part II*. Translated by Cohn, Paul V. Part of *The Complete Works of Friedrich Nietzsche*. Edited by Oscar Levy, The MacMillan Company, 1911, pp. 287–288.

fear. James also speaks of it as a peace driven by "mastery", that is, that kind of power that seeks to prevail rather than to reconcile. I am not sure what to make of Nietzsche's last remark here about the lightning coming from the clouds. But the image of lightning destroying in one fell swoop is clearly a martial one, of a violent overturning that does not in fact have patience for the kind of self-reflection that Nietzsche says our "liberal representatives" should engage in. How, then, should pacifism conceive of a peaceful way to peace?

Chapter 5

Disarming Power: Memory, Sacrifice, Grief, and Hope

The war against war is going to be no holiday excursion or camping party.[1]

These were the opening words of William James's essay "The Moral Equivalent of War" (1905). It is one of the most important essays in the history of pacifism because of the *challenge* it posed to pacifism. James framed that challenge in terms of a paradox of the pre-World War I era, of a kind that exists today. At the time, there was the widespread sentiment that war was an archaic institution, at odds with every standard of rationality. But this existed alongside a massive arms race and hosannas from many quarters about the glories of armed conflict. In James's day, the major European powers had 4 million men in uniform and the capacity to mobilize ten times that number. Today, there is also widespread disillusionment of war, but alongside resistance of the Western citizenry to participate in it. Yet the war system continues. "Everyone confesses 'war is horrible', yet we continue to have war" Stanley Hauerwas writes.[2] Circumstances seem to be excellent for creating a world beyond war. Yet that goal can seem as distant as ever.

Nonviolence's great voices have always stressed the need for persistence. The abolition of slavery has been a model. Voices from the Christian tradition have identified such persistence with the need for "left-handed power," a term they derive from Martin Luther's discussions of power. "Right-handed power" is the power of governments that threaten, coerce, and bully; it prides itself on its displays of power because its force requires spectacle. "Left-handed power" is the strength of the poor and the weak. It is *disarming power* in two respects: it disarms the power of the powerful, and it does so by a power that is disarming, that is, non-provocative,

---

1 James, William. *The Moral Equivalent of War*. Obscure Press, 2013, p. 3.
2 Hauerwas, Stanley. "Sacrificing the Sacrifices of War." *Journal of Religion, Conflict, and Peace*, 1(1), 2007. www.religionconflictpeace.org/volume-1-issue-1-fall-2007/sacrificing-sacrifices-war.

DOI: 10.4324/9781032686189-5

defusing of hostility. Scholars of slavery in traditional societies have evoked it to describe the more indirect and secretive forms of resistance that slaves, peasants, and servants have practiced.[3] It can designate more generally a power that is not afraid to show weakness for the sake of something greater; a power that grants freedom rather than impinges on it, and which acts in favor of community rather than hierarchy. Dr. King chastised the "Drum Major Instinct" for brazen gestures that cry for attention. Dorothy Day spoke of how a life devoted to peace is achieved "little by little."[4]

But preaching patience is not enough. James's challenge was how pacifism could become more expansive—hence more impactful. It was not enough to recount the case against war in quantitative terms in the manner of Norman Angell or Jan Gotlib Bloch. Pacifism needed to address the deeper issues raised by the human experience of war, including what J. Glenn Gray has termed its "enduring attractions."[5]

In this final chapter, I offer a set of reflections on some of the larger issues raised by the challenges of peace making and peace building today.

The end of the 20th century saw significant debates about how war should be memorialized, specifically how its *sacrifices* should be memorialized. Amidst the different proposals offered, the larger question raised was what it says about war that it is capable of provoking clashes over such elementary human matters as memory and sacrifice. War has always weaponized memory, but World War I and its progeny gave us something new— killing as a *mode* of remembering: we fight, not because we have any idea of the ultimate aim, but as a way of "honoring" those who have *already* fought and died. Note that this is another avenue to endless war, as each new death incurred in honoring the dead requires that it too be honored. And what it suggests is that a central task of disarming power is disarming *memory*. The question raised by the central place of war in the imaginary of the nation-state is whether community can be grounded in *another* form of sacrifice, or whether disarming sacrifice means doing away with sacrifice as the ground of community entirely. Do we need another *type* of war, as William James suggested, or should we seek a peace cleansed of war entirely? The prominence that the topic of grief has acquired in recent discussions of war is attributable, in my view, to the fact that grief *cannot* be weaponized, that grief's power is *already* disarming. We may kill as a way

---

3 Baptist, Edward E. *The Half Has Never Been Told: Slavery and the Making of American Capitalism.* Hachette UK, 2016, Kindle Locations 9447–9448; Scott, James, *Domination and the Arts of Resistance: Hidden Transcripts.* Yale University Press, 1990.

4 I have discussed these issues in Ryan, Cheyney, "The One Who Burns Herself for Peace." *Hypatia a Journal of Feminist Philosophy,* 9(2), 2009, pp. 21–39.

5 Gray, J. Glen. *The Warriors: Reflections on Men in Battle.* Bison Books, 1998, pp. 25–58.

of remembering but not as a way of grieving. Thus it offers a new mode of perception toward the violence of our time. This is why Jane Addams saw such promise in grieving war widows whose memories constituted a "challenge to war," and in so doing offered conditions for hope.

[I]/ Repentance Before the Flood?

A *New York Times* article of September 1, 1972, titled "US Aides in Vietnam See an Unending War" reported that, following a visit to Saigon of Henry Kissinger, "high ranking" American officials were "talking privately" of the possibility that the war would in fact never end.[6] I know that many of us at the time felt that the war would never end. The specter of endless war remains very much with us in today's talk of "forever wars"—referencing what I have termed war's inexorability.

A Biblical image of war is that of a *flood*. The surging flood does not just overpower its victims, it does so indiscriminately. Everything and everyone are washed away. The Biblical terms for a flood include one that evokes summary mass execution.[7] In the story of Noah, what precipitates the Flood is a condition of social "wateriness" that symbolizes the disappearance of the distinctly human capacities for morality and speech, both expressed in the fact that "the earth was filled with violence." The Noahic Covenant was the agreement God made with Noah after he and his family were spared from the Flood, an agreement sealed by a rainbow that has become the image of the bonds created by peace. Henceforth, the sacredness of life would be acknowledged and respected in the prohibition on taking human life. In return, there was the promise of no more floods—though if there was a next time, it would be by fire.

The flood's image of war as death-by-water is a recurring one. It is evoked in one of the most prominent antiwar songs of the Vietnam era by Pete Seeger. The song first became famous in 1967 when TV networks refused to let it be played. It has since become an antiwar anthem throughout the world, including Russia.[8]

The song describes an ill-fated platoon mission of World War II in which their captain insists that his soldiers forge on into an ever-deeper bog that

6 Whitney, Craig R. "U.S. Protests Invective by Saigon on McGovern." *The New York Times*, September 11, 1972. www.nytimes.com/1972/09/11/archives/us-protests-invective-by-saigon-on-mcgovern.html.

7 See Zornberg, Avivah Gottlieb. *The Beginning of Desire: Reflections on Genesis*. Random House LLC, 2011.

8 The song was translated into Russian by Alexander Dolsky, who performed the song in concerts in the 1980s during the Soviet–Afghan War, which has been referred to as the "Soviet Union's Vietnam."

threatens to consume them all. But the tale is not about World War II. It is really about Vietnam and the futility of war generally.[9] In the song, the captain leading the platoon ignores the protests of his sergeant, berating him as a "Nervous Nellie," as they sink deeper and deeper into the mud. The platoon is only spared when the captain sinks completely into the muck, leaving only his helmet to float by.

The song—"Waste Deep in the Big Muddy"—concludes:

Now I'm not going to point any moral—
I'll leave that for yourself.
Maybe you're still walking, you're still talking,
You'd like to keep your health.
But every time I read the papers, that old feeling comes on,
We're waist deep in the Big Muddy
And the big fool says to push on.

Here, the danger is obvious to everyone except the officer leading. Everyone can see the fate that awaits them. But an issue I raised in the last chapter is that the dangers of war building are both deeper and less obvious. What they evoke is not a river sweeping us away but something more chthonic: something seeping silently but relentlessly beneath the ground—and all the more ominous for that fact.

In the distance from my apartment in Portland, Oregon, I can see the Columbia River as it widens to empty into the Pacific Ocean. The river is a magnificent national treasure, celebrated in Woody Guthrie's "Roll On, Columbia, Roll-On." If one journeys up the Columbia River, one will encounter the Hanford Nuclear Reservation, arguably the most polluted area in the United States. This is a headline of a New York Times article that appeared as I was writing this:

A POISONOUS COLD WAR LEGACY THAT DEFIES A SOLUTION[10]

Hanford was once the nation's principal site for processing plutonium for nuclear weapons. It was vital to the Cold War when America produced an average of four nuclear bombs a day, 365 days a year. It is huge, larger than Los Angeles in square miles. The camp that was built to house its

---

9 The story is reminiscent of an actual event termed the "Ribbon Creek incident" in which a Marine drill instructor marched his men into a swampy tidal creek drowning six of them.

10 Vartabedian, Ralph. "A Poisonous Cold War Legacy That Defies a Solution." *The New York Times*, May 31, 2023. www.nytimes.com/2023/05/31/us/nuclear-waste-cleanup.html.

workers during World War II is still the largest in American history. Hanford housed the nation's first full-scale nuclear reactor, and its nuclear plant for producing electricity was the largest power reactor in the world well into the 1960s.

It was a total secret in World War II. Then, and for decades afterward, those who worked in and lived around it were not told what really went on there. They were ignorant of what was causing their higher incidents of thyroid, reproductive, and nervous system tumors that were later linked to radiation exposure. The facility is still run as a covert military operation—which is why few Americans know about it even now. I only know about it because I have lived in the Northwest and learned of it when it became an object of protest several decades ago.

The sign at Hanford's entrance off interstate Highway 90 reads: "WELCOME TO HANFORD—WHERE SAFETY COMES FIRST—ENVIRONMENTAL EXCELLENCE." It houses 54 million gallons of radioactive bomb-making sludge that according to the New York Times they still do not know what to do with. Hanford is not alone. Similar plants in South Carolina, Ohio, and Idaho have tons of debris that will be radioactive for thousands of years; two million pounds of mercury remain in the soils and waters in eastern Tennessee. Years ago, 177 auditorium-size underground tanks were built to house Hanford's radioactive waste. With a life span of about 20 years, the assumption was that they would devise a better solution in time. They have not. Engineers thought they had solved the problem with a plan to pump the sludge into impenetrable glass tanks and bury them in the desert. But construction of the five-story chemical treatment plant for doing this was halted (after the expenditure of $4 billion) due to unsolvable safety defects. Alternative plans that are now being developed presume it will never be fully cleaned up. Hundreds of thousands of gallons of toxic waste will be left.

The unfinished treatment plant still stands as a symbol of war building's shrouded menace. Radioactive waste remains in shallow underground tanks that are certain to fail long before the waste ceases being radioactive. All of this is just a few miles from the Columbia River, a vital lifeline for cities, farms, Native American tribes, and wildlife throughout the Northwest. The state of Washington has estimated that about 300 gallons of radioactive sludge per year are seeping into the soil.

There are Hollywood movies about the Manhattan Project. The story lends itself to drama, involving as it did feats of scientific and engineering brilliance by young people all cooped up in one place. Manhattan Project-movies are not battle movies, but the story of the atomic bomb's creation ends with an awe-inspiring explosion that could be seen from the farthest distances. There are no movies about Hanford, nor will there ever be. Manufacturing nuclear material is too prosaic and too much happens underground. But if a movie could be made, it would portray the threat of war building at its most malevolent.

How do we mobilize people against war's silent sludge?

Idealism is often dismissed as the privilege of those who have never fully encountered the real world. "No one would entertain such fantasies of peace if they knew what war was really like." But some of the most hopeful visions for ending war have emerged from the midst of war and from soldiers themselves. It was the radicalism of disgruntled soldiers in the 17th-century English Civil War that gave birth to some of the first doctrines of human rights. Some of those soldiers went on to found pacifist sects like the Quakers.

Altiero Spinelli (1907–1986) was an Italian former communist and militant antifascist who spent ten years imprisoned by Mussolini's fascist regime. After the war, he became a central figure in projects for European unity. He is often referred to as one of the founding fathers of the European Union. While in prison, he and fellow democratic socialists in 1941 drafted the *Manifesto for a Free and United Europe*—most commonly known as the *Ventotene Manifesto*.[11] Because of the need for secrecy and a lack of proper materials, the *Manifesto* was written on cigarette papers and smuggled out in the false bottom of a tin box, then it was widely circulated in resistance movements throughout Europe. It was a formative document when resistance leaders including Spinelli met clandestinely in Geneva in 1944.

The urgent question for those who experienced the horrors of World War II was: how can we create a world beyond war?

For the *Manifesto*, the issue was nothing less than the "crisis of modern civilization" itself. At the heart of that crisis was civilization's penchant for perpetual war. Its world was one in which "Repeated wars force men to abandon families, jobs, property, and even lay down their lives for goals, the value of which no one really understands." Indeed, the "goals" of these wars were essentially delusions, of which two stood out. One was the idea of "*race*."

> Although nobody knows what a race is, and the most elementary understanding of history brings home the absurdity of the statement, physiologists are asked to believe, demonstrate and even persuade us that people belong to a chosen race, merely because imperialism needs this myth to stir the masses to hate and pride.

---

11 Spinelli, Altiero. "For a Free and United Europe. A Draft Manifesto;" otherwise titled, "The Manifesto of Ventotene." English Translation, Union of European Federalists, 1941. www.federalists.eu/uef/library/books/the-ventotene-manifesto/.

The other was the idea of "*living space*." This originated in a "pseudo-science of geopolitics [that] has been created in an attempt to prove the soundness of theories about living space and to provide a theoretical cloak to the imperialist desire to dominate."

The *Manifesto* insisted that, even if World War II's fight against fascism was successful, it would have been for nought if it did not uproot the factors that made for perpetual war. So what was needed was a revolutionary project "bold enough to criticize the old political approaches." Lacking this, it would just be a matter of time until the forces of change become "ensnared" in the same old practices—and war returned again.

What was needed was a "complete reorganization of society" to transcend the war-prone arrangements of nation-states and empires. The *Manifesto* acknowledged that the ideology of national independence had been a powerful stimulus to progress. "It helped overcome narrow-minded parochialism and created a much wider feeling of solidarity against foreign oppression." It had provided a critique of empire, and sometimes a critique of states. But now the nation had become an idol.

> It has changed to become a divine entity, an organism which must only consider its own existence, its own development, without the least regard for the damage that others may suffer from this. The absolute sovereignty of national States has led to the desire of each of them to dominate, since each feels threatened by the strength of the others . . . This desire to dominate cannot be placated except by the hegemony of the strongest State over all the others.

If enduring peace is to be achieved, then the goal could only be "the definitive abolition of the division of Europe into national, sovereign states."

But here is the problem: the most significant attempts to address war have only come *after* major wars. Philip Bobbitt has recounted how large-scale changes in global society have all been a consequence of (what he calls) epic wars. Plus the call for radical rethinking after such wars is often short lived. Consider the case of nuclear weapons in the immediate aftermath of Hiroshima and Nagasaki.

Lewis Mumford was one of mid-20th-century America's major public intellectuals. In the late 1930s, he graced the cover of Time Magazine for his writings on urbanism that appeared in mainstream publications and popular books. He was a quiet, unassuming, bookish man almost always photographed in the drab gray suit of an accountant. Now, Mumford is totally forgotten except in urban studies. He rarely receives mention in writings on nuclear weapons and their dangers. I think his disappearance is due to the heartfelt alarm he voiced about the specter of nuclear weapons and what they signified about the corruption of society generally. He was deeply influenced by Randolph Bourne.

Just six months after Hiroshima and Nagasaki, Mumford published an article in the *Saturday Review of Literature* titled "Gentlemen: You are Mad!."[12] While others celebrated the war's end, Mumford saw it as inaugurating a new phase of insanity. He did not mince words:

> The madmen are planning the end of the world. What they call continued progress in atomic warfare means universal extermination, and what they call national security is organized suicide. There is only one duty for the moment: every other task is a dream and a mockery. Stop the atomic bomb. Stop making the bomb. Abandon the bomb completely. Dismantle every existing bomb.

Mumford thundered for the rest of his life about nuclear weapons and the maladies of the war system that they exemplified. "As soon as nuclear weapons were invented, the irrationality of war became total." The problem was technological, specifically humanity's fixation on the kind of technology that idolized "the machine" above all else. "In our one-sided preoccupation with science and technics, we have taken the automatic machine as the model for human behavior." This has led to the greatest paradoxes.

"Our age has been characterized as the Nuclear Age or the Space Age, to celebrate modern man's sudden command of time, space, and energy, in a fashion more absolute than he ever before dared to dream. But this bright side of science and invention has a dark face. As never before, ours has proved to be an age of mass extermination and mass destruction."

And there is no technological solution.

"There is no mechanical escape from the dangers mankind as a whole now confronts: no rocket missiles or counter-missiles will save us. We must find the human way out. Instead of planting our flag on the moon, we must quickly move to reclaim this planet for humanity."

This is because the problem is a moral one, reflecting our attempts to apply "a savage stone age morality to an atomic age civilization." "What should cause the deepest concern at this moment is not our technological but our moral backwardness." That backwardness involves "delusions" of infallibility that come with worshiping the powers of the machine, whereas "To be human is to admit that we are limited and fallible, frail in all our powers, and prone to perpetual self-deception, fomented by pride." In the

---

12 Mumford, Lewis. "Gentlemen: You Are Mad!" *Saturday Review of Literature*, March 2, 1946, pp. 5–6.

end, Mumford's prophetic-type pronouncements intimated that any solution must be a *spiritual* one. "Our behavior is explainable only on the assumption that a whole side of our consciousness has been cut off from contact with reality, by a sort of moral nerve-block." It is "this very insulation that keeps us from taking the measures necessary to correct our errors: above all, it keeps us from taking the first step, and this is to repent."[13]

Talk of "repentance" sits uneasily with many of us. It is worth remembering that its original meaning is that of a turn, or leap, from being a mere observer to an active participant in preventing what jeremiads like Mumford's perceive as a looming catastrophe. It returns us to the question:

How do we engage people to repent *before* the flood?

This has been the challenge faced by the environmental movement, though contemporary events are now placing floods (and fires, and other climate change calamities) before our eyes. Some have suggested that the problem is not the absence of crisis but the presence of a new *kind* of crisis, of slow not always visible threats but ones that are merging together. This is the view of those who insist that the challenge of peace be framed as part of the larger challenge of security. But there are specific problems raised by the challenge of war that impact our capacity for responding in a truly human way to it.

Let me return, then, to the issues of memory and sacrifice noted at the start.

[II]/ Excitement and Sacrifice

William James regarded the "paradox" of his time as taking a specific form. I quote him in full.

There is something highly paradoxical in the modern man's relation to war. Ask all our millions, north and south, whether they would vote now (were such a thing possible) to have our war for the Union expunged from history, and the record of a peaceful transition to the present time substituted for that of its marches and battles, and probably hardly a handful of eccentrics would say yes. Those ancestors, those efforts, those memories and legends are the most ideal part of what we now own together, a sacred spiritual possession worth more than all the blood poured out. Yet ask those same people whether they would be willing in cold blood to start another civil war now to gain another

13 All of these quotes are from Mumford, *Op. cit.*

similar possession, and not one man or woman would vote for the proposition. In modern eyes, precious though wars may be, they must not be waged solely for the sake of the ideal harvest. Only when forced upon one, only when an enemy's injustice leaves us no alternative, is a war now thought permissible.[14]

The reference to "our war for the Union" is to the American Civil War (1860–1865) which had concluded 40 years before this essay. That war was roughly the same distance from James as the Vietnam War is from Americans today. Like the Vietnam War, it posed the question of memory, of what should be kept and what should be "expunged" from our history.

What is the "paradox" it posed?

The first way to take James's point is that, while his contemporaries liked the *idea* of the Civil War and all its glories, they had no inclination to ever fight in such a war *themselves*. But this would be contrary to the concern of his essay. His concern was not why people refused to go to war but why they *did* go to war. Rather, I take his point to be about justification—that people did not regard the glories of past wars as sufficient *justification* for going to war, that is, the only legitimate justification was when wars were forced upon us; yet they still regarded such glories as rendering war permissible. For James, the issue involved the limits of justification in matters of war, or the limits of rationality generally.

There are two ways to take his point here. The first appeals to the *experience* of war as one that is exciting, to the point of *intoxicating*—so that questions of "justification" are irrelevant to whether people are drawn to engaging it. Pacifists, we might say, are Apollonian while war is Dionysian. Thus construed, James's point is to suggest that pacifism will not get anywhere until it provides an *alternative* form of excitement, which James feels could only be provided in an alternative form of war. Hence, his proposal that war between societies be replaced by a collective "war against nature."

The theme of war-as-exciting is one of the more persistent tropes in the literature of war. A notable recent example is Chris Hedges' *War Is a Force That Gives Us Meaning*.[15] Hedges writes of his first-hand experiences as a war correspondent that,

> The lust for violence, the freedom to eradicate the world around them, even human lives, is seductive. And the lines that divide us, who would like to see ourselves as civilized and compassionate, from such

---

14 James, William. *The Moral Equivalent of War. Op. cit.*, p. 3.
15 Hedges, Chris. *War Is a Force That Gives Us Meaning.* Public Affairs, 2002.

communal barbarity is razor-thin. In wartime it often seems to matter little where one came from or how well schooled and moral one was before the war began. The frenzy of the crowd is overpowering.[16]

Bringing his experiences to bear on American society immediately after 9/11, Hedges writes. "We believe in the nobility and self-sacrifice demanded by war, especially when we are blinded by the narcotic of war. We discover in the communal struggle, the shared sense of meaning and purpose, a cause. War fills our spiritual void."[17] And because of this, "We no longer seem chastened by war as we were in the years after the Vietnam War."[18]

Undoubtedly, people find the *idea* of war exciting, though some of James's remarks suggest that this is distinct from finding the actual experience of it exciting. Indeed, the idea may mainly be exciting to those who have *not* experienced it. In his 1904 talk to the World's Peace Congress, James suggested that war was the "supremely thrilling excitement" "*especially for non-combatants.*"[19]

The remotest spectators share the fascination of that awful struggle now in process on the confines of the world. There is not a man in this room, I suppose, who doesn't buy both an evening and a morning paper, and first of all pounce on the war column. A deadly listlessness would come over most men's imagination of the future if they could seriously be brought to believe that never again *in soecula soeculorum* would a war trouble human history.

But what—precisely—follows from this? "The Titanic" was once the most popular movie of all time. It revolved around an older woman for whom that disaster was the peak experience of her life. It was not just exciting but provided the occasion for the heroism of both her and her lover. But this says nothing about the actual experience of shipwrecks, or the "attraction" of actual shipwrecks, or the "value" of actual shipwrecks as occasions for heroism, etc.

Is the actual experience of wars exciting, explaining why people do not just go to movies about them but flock to fight in them?

As a generalization, I have always found this to be nonsense. I come from a generation, of the Vietnam War, whose members almost universally

16 *Ibid*. 172.
17 *Ibid*. 158.
18 *Ibid d*, 160.
19 James, William. "Remarks at the Peace Banquet." *Memories and Studies*. Longman Green and Co., 1911, pp. 299–306.

did everything they could to *avoid* going to war. I sat through endless deliberations with my contemporaries over whether to go to Vietnam. I do not remember a single one of them ever saying, "on the other hand, fighting in Vietnam would be exciting!" Moreover, if young men found war to be so "seductive," then why has mobilizing them in large numbers always required conscription—perhaps the most coercive institution of modern society? Randolph Bourne raised this point.

> Men are told simultaneously that they will enter the military establishment of their own volition, as their splendid sacrifice for their country's welfare, and that if they do not enter they will be hunted down and punished with the most horrid penalties; and under a most indescribable confusion of democratic pride and personal fear they submit to the destruction of their livelihood if not their lives, in a way that would formerly have seemed to them so obnoxious as to be incredible.[20]

Yes, there are *some* young men who have found war exciting. Having been a young man myself, I can attest to the fact that young men find all sorts of idiotic things exciting. But this does not mean that societies should organize themselves around them.

A second way to take James's point is that war has a value *beyond* justification. Some have invoked war's aesthetic value; Kant waxed eloquent about its sublimity. Another variation is that war has a value *prior* to justification insofar as it is the founding act of the community—which provides the conditions of any justification.

This is a thought we find in Hegel and in some of his remarks on heroism. "The heroes who founded states, introduced marriage and agriculture, did not do this as their recognized right, and their conduct still has the appearance of being their particular will," he writes in his *Philosophy of Right*. If heroism involves "right" at all, it is "the right of heroes to found states." Society cannot be founded on contractual type relations alone; it needs war—and specifically, the *sacrifices* of war—to create and sustain it. It is only *after* the state is founded that talk of what is "justified by right" makes sense. Heroes come on the scene

> only in uncivilized conditions. Their aim is right, necessary, and political, and this they pursue as their own affair. But as the higher right of the Idea against nature, this heroic coercion is a rightful coercion. Mere goodness can achieve little against the power of nature.[21]

20 Bourne, *Op. cit.*, p. 12.
21 Hegel, G.W. *Hegel's Philosophy of Right*. 1820. Translated by T. M. Knox, Clarendon Press, 1952, paragraph 93.

My account of nationalism has postulated an intimate relation between the sacrifices of war and the national community, grounded in an imperative to honor the sacrifices. A reference point for someone of James's generation would have been Abraham Lincoln's "Gettysburg Address," the founding text of American nationhood. For Lincoln, the national community was founded on a debt to the dead whose sacrifices had served to "consecrate" the land on which they died, obliging us to periodically "renew" that bond through our own sacrifices.[22] In his homage to Lincoln, Walt Whitman made the bond between national community and death explicit:

> The final use of a heroic eminent life—especially of a heroic eminent death—is its indirect filtering into the nation and the race, and to give . . . a cement to the whole people, subtler, more underlying, and anything in written constitutions, or courts or armies . . . Namely, the cement of a death identified thoroughly with that people. It has the greatest power of all to condense debt that perhaps only really lastingly condensed . . . A nation.[23]

In sum, the values of war are such for James that pacifists will get nowhere until they can provide an alternative form of war and its sacrifices. Pursuing this point further would mean interrogating the picture of "sacrifice" presumed by war and of the social "bonds" constituted by such sacrifices. These are among the concerns that I find addressed by Jane Addams, for whom the questions of memory and sacrifice could not be detached from the question of gender and how it informed our conception of grief.

[III]/ Alcohol, Heart-Broken Women, and Hope

I take Jane Addams as raising the question of how war weaponizes memory and sacrifice, and as suggesting that doing so leaves no place for real grief. A truly expansive form of pacifism would be one that speaks from that grief. Its perspective would not be that of the disengaged observer but of one who reaches down into the "valley of sorrow that is war." I have said much about the 20th century in this work. The next step may be learning how to grieve it.

---

22 See Marvin, Carolyn and Ingle, David. *Blood Sacrifice and the Nation: Totem Rituals and the American Flag.* Cambridge University Press, 1999.
23 Whitman, Walt. "Death of Abraham Lincoln Lecture." *Complete Works of Walt Whitman.* Delphi Classics. 2012, p. 1949.

[1]/ Isolation and Despair

Let me return to the story of Addams that I introduced in Chapter 2. Jane Addams was the most admired woman in America when World War I began. One year she ranked higher in public opinion than the Virgin Mary. I have noted the truly heroic efforts that she and other feminists made early in the war to press for a negotiated settlement, traveling to European capitals even amidst the fighting. But then she came back to America.

Upon returning to New York, Addams gave a speech at Carnegie Hall recounting her experiences. In the course of her talk, she remarked offhandedly on how soldiers were supplied with alcohol prior to major battles to keep their courage up. She explained,

> The young men in these warring countries say, "Ah, the bayonet charge, that is what we dread!" You know, they make their men drunk before they can get them to charge. They give them beer in Germany, rum in England, and absinthe in France. They have to give them dope, as the saying is, before a bayonet charge as possible.[24]

Keep in mind: Addams was speaking two years *before* America entered the war, and she was talking about foreign soldiers whose cause was regarded by many Americans with deep ambivalence. Still, her remarks created a firestorm.

She was immediately vilified in the popular press, most notably by Richard Harding Davis, America's leading war correspondent and previously a friend of Addams's. Davis exclaimed,

> In this war the French or English soldier who had been killed in a bayonet charge gave his life to protect his home and country. For his supreme exit he had prepared himself by months of discipline. Through the winter in the trenches he had endured shells, disease, snow and ice. For months he had been separated from his wife, children, friends—all those he most loved. When the order to charge came it was for them he gave his life that against those who destroyed Belgium they might preserve their home, might live to enjoy peace. Miss Addams denies him the credit of his sacrifice. She strips him of honor and courage. She tells his children, "Your father did not die for France, or for England, or for you; he died because he was drunk." In my opinion, since the war began, no statement has been so unworthy or so untrue and ridiculous.

---

24 See McAndrew, Tara. "Local Icon Shifts To 'The Most Dangerous Woman in America'." *Illinois Public Media*, July 11, 2017.

The contempt it shows for the memory of the dead is appalling; the crudity and ignorance it displays are inconceivable.[25]

The role of alcohol in war was well known to anyone familiar with war and had been widely reported in Europe. Winston Churchill once attributed the fighting spirit of the British Navy to "Rum, sodomy, and the lash." But Addams's reputation suffered a catastrophic collapse from which it never fully recovered. She regarded her peace activism as her principal achievement, but it has been largely forgotten today even by many in the peace community.

She was truly shocked by the vehemence of the response against her. "I was conscious. . . . That the story had struck athwart the popular and long cherished conception of the nobility and heroism of the soldier as such."[26] She was perceived as challenging the ideology of death-by-sacrifice at the heart of all nations. She was not just an enemy to her community. She had placed herself against community itself as grounded in war. Addams did not discount the heroism to be found in war. But she did not, like James, seek another kind of war to display martial sacrifice. Rather, she looked to other forms of sacrifice in war that she construed as standing *against* war. She found them in the experiences of older women.

The idea that women represent an alternative form of memorialization to war is as old as Antigone. Addams quoted Romain Rolland's call: "Be a living peace in the midst of war—the eternal Antigone refusing to give herself up to hatred and knowing no distinction between her suffering brothers who make war on each other." Addams's focus was the mothers she had encountered during her European peace-making efforts, those "brokenhearted women" already widowed and rendered childless by war: "Desolated women, stripped by war of all their warm domestic interests and of children long cherished in affectionate solicitude, [who] sat shelterless in the devastating glare of Memory."[27]

Her story of these women, *The Long Road of Woman's Memory*, was composed during her own time of isolation from speaking out against the war. The isolation of the elderly had already been a theme of hers in her writings on the communities she worked with in Chicago. In *The Long Road of Woman's Memory*'s final chapter, "Women's Memories— Challenging War," she seemed to find in the "abominable and constant loneliness, which is almost unendurable" (in the words of one war widow)

---

25 Harding Davis, Richard, letter to the *New York Times*, July, 13, 1915.
26 Addams, Jane. *Peace and Bread in Time of War*, Op. cit., p. 79.
27 Addams, Jane. *The Long Road of Woman's Memory*. Sagwan Press, 2015, p. 6.

a standpoint to resist war.[28] She was surprised by this. She had expected to find that memories from loneliness would play a consoling role, that they would serve to reconcile the elderly with their fate. What she found instead was, in her words, that the

> reminiscences of the aged, even while softening the harsh realities of the past, exercised a vital power of selection which often necessitates an onset against the very traditions and conventions commonly believed to find their stronghold in the minds of elderly people.

Marx held that the proletariat's outcast status would eventually provide it with the perspective to pierce the assumptions and conventions of the established order. Addams finds the same potential in those that have been completely disinherited by war.

> In its various manifestations the struggle in women's souls suggests one of those movements through which, at long historic intervals, the human spirit has apparently led a revolt against itself, as it were, exhibiting a moral abhorrence for certain cherished customs which, up to that time, had been its finest expression.

The "cherished customs" she has in mind are specifically those of *sacrifice*. Slaying one's own child, she writes, used to be identified with "piety," "courage," and "devotion to ideals." Now these are identified with sacrifice in war and with the sanctity of the community that is grounded in such sacrifice. She takes rejecting all this to be the particular burden of the "suffering mothers of the disinherited" whose call is for justice as well as peace, a call "to defend those at the bottom of society who, irrespective of the victory or defeat of any army, are ever oppressed and overburdened."[29]

"I have never been a Feminist," she reports one woman saying, "but during the last dreadful months, in spite of women's widespread enthusiasm for the war and their patriotic eagerness to make the supreme sacrifice, I have become conscious of an unalterable cleavage between Militarism and Feminism." What is it about their grief that reveals to them the "great essentials," in Addams's words, of the "common human experience"? I think an answer to this could be a major contribution to what might be termed an epistemology of nonviolence. The standpoint of the

---

28 *Ibid.*, p. 62.

29 For more discussion of war and grief, see my "The Lament of the Demobilized." *To End a War: Essays on Justice, Peace, and Repair*. Edited by Graham Parsons and Mark Wilson. Cambridge University Press, 2023.

outcast, of the isolated, of the lonely is akin to that of the outcast prophet, whose message has been portrayed as one of awakening the community to its grief. The problem, again, is one of *power*: the threat of grief to the established order is the threat of that which cannot be controlled, however much societies try to do this by entombing grief in "war memorials." Empires especially are incapable of listening and responding. In his eloquent *The Prophetic Imagination*, Walter Bruggeman writes that

> Only in the empire are we pressed and urged and invited to pretend that things are all right. . . . And as long as the empire can keep the pretense alive that things are all right, there will be no real grieving and no serious criticism

hence "Bringing hurt to public expression is an important first step in the dismantling criticism that permits a new reality, theological and social, to emerge." He finds in this a principle of motivation.

> If the task of prophecy is to empower people to engage in history, then it means evoking cries that expect answers, learning to address them where they will be taken seriously, and ceasing to look to the numbed and dull empire that never intended to answer in the first place.[30]

The talk here of "empowering people to engage in history" returns us to the question of repentance. In *The Meaning of Revelation*, H. Richard Niebuhr identified repentance with a radical reorientation of perspective, which he identified with two different stances on history.[31] One is the view from "without," the impersonal perspective that seeks objectivity; the other is the view from "within," the personal perspective that seeks meaning. This puts things crudely, of course, but for Niebuhr the importance of approaching history from within was that only if one does so is one moved to embrace it as one's own concern in a way that compels one to act. But there is no mechanical way to move from one to the other. It involves what seems like a Kierkegaardian leap, but it is available to everyone in their capacity to hear what history says to them if they will open themselves to it. Bruggeman puts this as hearing the cries that expect answers. Dorothee Soelle identifies this movement as one from helplessness to power that is

---

30 Brueggemann, Walter. *The Prophetic Imagination.* Fortress Press, 2001, pp. 11–13.

31 Niebuhr, H. Richard. *The Meaning of Revelation.* Macmillan, 1941. Cited in Yoder, John Howard. *Preface to Theology: Christology and Theological Method.* Baker Publishing Group, 2002.

provoked by the public expressions of suffering as protest.[32] The suggestion is that we no longer require new crises, hence more war, for this to be achieved. But we do require the capacity for hope.

[2]/ Hope and Love

The French philosopher Gabriel Marcel (1889–1973) is largely forgotten now in Anglophone philosophy. He is among those figures whose philosophical voice was deeply impacted by his experiences in World War I. As a noncombatant with the Red Cross, his job was obtaining information about wounded and missing soldiers and then informing their relatives of their fate. We might say that his job placed him in the very midst of sorrow and grief. He subsequently observed that these intense personal encounters left him with a lifelong philosophical suspicion of what he called the "spirit of abstraction." It was during World War I that Marcel began his *Metaphysical Journal* that anticipated much of his later thinking. While disavowing what he termed "dogmatic pacifism," Marcel acknowledged that war's "desolate aspect" rendered it an "object of indignation" that was a "horror without equal" which philosophical thinking was obliged to confront.[33] What made it such an evil was that war required hope to ever transcend it, but war's impulse was to kill hope as part of killing everything that is precious to our humanity.

Marcel returned to this theme explicitly in his post-World War II book, *Homo Viator.*[34] The crucial distinction for him was between optimism and hope. He sees optimism as a confidence that things will get better, a confidence based on objective evidence, where "objective" means deduced from factors in which the optimist is not themselves involved. It has a spectral dimension, then, allowing for a posture of passivity. Surveying the omnipresence of war, it is hard to be optimistic.

Hope, by contrast, is fundamentally creative in how it goes beyond the facts, specifically in how it sees more possibilities in a situation than the facts may themselves admit. But even this may put it too spectrally, as the possibilities are as much acted on as perceived insofar as hope takes the possibilities it envisions as ones to be acted on now. In Roberto Unger's words, "It instantiates a conceived future rather than merely looking to

32 Soelle, Dorothee. *Suffering.* Translated by E. R. Kalin, Fortress Press, 1975. See also Soelle, Dorothee. *Revolutionary Patience.* Wipf and Stock, 2003.
33 Marcel, Gabriel. "Autobiographical Essay." *The Philosophy of Gabriel Marcel: The Library of Living Philosophers*, 17. Edited by Paul Arthur Schilpp and Lewis Edwin Hahn. Open Court, 1984, p. 20.
34 Marcel, Gabriel. *Homo Viator: Introduction to a Metaphysic of Hope.* Henry Regency, 1951.

it."[35] This is what it means to perceive pacifism as fundamentally a response, both theoretical and practical. And it is what it means to regard pacifism, as its major figures from Tolstoy to Gandhi to Dr. King to Dorothy Day and on have regarded it, as fundamentally informed by love. In the old Quaker saying, it is "To see what love can do." For what else is love but the creative capacity to envision more possibilities that might be obvious and to nurture their achievement by enacting them now?

35 Unger, Mangabeira Roberto. *Passion: An Essay on Personality*. Simon and Schuster, 1986, p. 245.

Appendix

Pacifism and Self-Defense

The justification of wars of self-defense is very convenient, since so far as I know there has never yet been a war which was not one of self-defense.

—Bertrand Russell[1]

In times of need—and politics are chronically in a time of need—the rulers were always able to evoke "exceptional circumstances," which demanded exceptional measures of defense. Since the existence of nations and classes, they live in a permanent state of mutual self-defense, which forces them to defer to another time the putting into practice of humanism.

—Arthur Koestler[2]

I have suggested that the challenge to pacifism—"What about self-defense?"—is often raised not as an invitation to seriously discuss the issue but as a way of rejecting pacifism out of hand. But pacifists have been equally dismissive of the issue. Most pacifists have felt that, in reality, war has as much to do with personal self-defense as rape has to do with procreation. Or take Tolstoy's attitude. When Tolstoy was challenged whether he would commit violence to prevent the murder of a child, he replied that, in his 75 years, he had heard of no one who actually faced such a dilemma, while "I perpetually did and do see not one but millions of brigands using violence toward children and women and men and old people and all the labourers, in the name of a recognized right to do violence to their fellows." These rejoinders are serious ones, but there is more to the issue. This appendix addresses some of the questions raised by self-defense for pacifism.

1 Russell, Bertrand. "The Ethics of War." *The International Journal of Ethics*, 25(2), 1915, pp. 127–142.
2 Koestler, Arthur. *Darkness at Noon*. Penguin, 1987, p. 128.

Let me begin with the general point that, despite self-defense's importance for contemporary international law, I think the crucial question is not self-defense but *security*. These are closely related but they are not the same thing; indeed, as Henry Shue and Jeremy Waldron have noted, "security" remains a relatively unexplored concept compared to the extensive analyses of self-defense—and yet how best to achieve security is in my view the *crux* of the dispute between war-ism and pacifism.[3] This is not an issue to be resolved abstractly, much less simply. The pacifist claim is that the war system has left us increasingly *insecure* as a global community and that security can be better achieved via nonviolent methods and arrangements. No one confronts the pacifist with the question "What about security?" because it is too obvious that addressing this question requires a serious discussion which for the pacifist would begin with addressing the tenuous world that war has created.

The appeal to self-defense is taken to have special force against pacifism because the claims of self-defense and their relation to war are taken to be simple and straightforward. Pacifism is taken to challenge what no one could possibly doubt. But self-defense has in fact played a complicated and changing role in relation to war. The recent discussions of self-defense in just war theory have revealed some of the philosophical complexities involved.

[I]/ Historical Complexities

At the start of the just war tradition, Augustine grounded it on a *rejection* of self-defense, in the following sense. Jesus had called on Christians to turn the other cheek, which early thinkers like Tertullian took to be a flat rejection of war. ("The Lord in disarming Peter henceforth disarms every soldier."[4]) Augustine took this as referring to the *spirit* in which violence

---

3 A shift from the right to self-defense to the right to security is more substantial than might first seem. They are different *kinds* of rights: the right to self-defense is the right to *act* in a certain way; this is why it is appealing as a starting point for thinking about war—because it seems clear, crisp, and straightforward. The right to security is the right to a *condition*. It does not lend itself to analysis in terms of abstract hypothetical cases, as self-defense does. Perhaps because of this, security and the right to security have received scant attention in political philosophy. Jeremy Waldron notes this, in his recent explorations of the notion. "It is shocking to discover how little attention has been paid to the topic of security by political philosophers". (Waldron, Jeremy. *Torture, Terror, and Trade-Offs: Philosophy for the White House*. Oxford University Press, 2010). See also Shue, Henry. *Basic Rights: Subsistence, Affluence and U.S. Foreign Policy*. Princeton University Press, 1996.

4 Tertullian, Quintus. *De Idololatria (On Idolatry)*. Translated by Rev. S Thelwall, 2015, Chapter XIX. www.documentacatholicaomnia.eu/03d/0160-0220,_Tertullianus,_De_ Idolatria_[Schaff],_EN.pdf.

was done, and as meaning that violence should not be inflicted purely for oneself. But violence was permissible in defense of others and in defense of those worldly arrangements necessary to protect others. As an act of altruism, soldiering was a matter of self-sacrifice involving a sacrifice of one's moral purity such that taking life remained a sin requiring penance afterward. Later ethics like chivalry continued this theme of war as defense of the weak, linking it with pagan virtues of honor.

The importance of self-defense as a personal right arose in early modernity with the individual rights framework generally. Privileging the right to self-defense was championed by the Humanists against the Scholastics as an alternative to endless theological debates and the conflicts they engendered. It was something that everyone might agree with.[5] Yet despite its centrality for figures like Hobbes, self-defense remained marginal as a consideration in wars between European powers. They were conceived as more like civil disputes in law, specifically disputes over rightful possession of property, often related to issues of inheritance. The question of self-defense, as the question of "who-hit-who first," was of no more consequence than who sued who first in a civil dispute. All that mattered was a fair resolution of the dispute. States had a right to defend themselves just as parties in a civil suit have a right to defend themselves; but in neither case was self-defense what wars are primarily *about*. And, as in a civil suit, which side would prevail was rarely a question of which side would *survive*. Self-defense, that is, was not a life-and-death matter.

It was in conflicts outside of Europe that self-defense was most construed as a life-and-death matter, as a matter of survival. The Crusades introduced the notion of defending the faith against the heathens, which was then transported back to Europe during the religious wars. In the 18th and 19th centuries, self-defense was principally invoked to justify European wars against Indigenous peoples. This is how Andrew Jackson and Euro-Americans portrayed the genocide of Native Americans. The Napoleonic Wars were sometimes framed as ones of survival, for example, the survival of traditional forms like the dynastic state against the revolutionary tide. But self-defense only came into its own in the late 19th century with the rise of nationalism, especially of the ethnic sort, which conceived of nations as quasi—biological organisms that according to Social Darwinism were engaged in a "survival of the fittest," with the caveat that "survival" required expansion such that those incapable of expanding would perish.

This came to a head in World War I where all sides framed the war as one of "survival" now identified with "races" like the "Huns,"

5 Tuck, Richard. *Natural Rights Theories: Their Origin and Development*. Cambridge University Press, 1982.

"Anglo-Saxons," etc.—each of whom carried the mantle of "civilization." German academics pronounced that if they lost "the world loses its light and its home of justice." The gendered dimension was crucial. Hew Strachan notes that the appeal to self-defense was essential to mobilizing the support of women, otherwise "more dubious about the war than men." If a woman's support for the war could be seen as an act of maternal responsibility, then "a primitive and basic response could be rolled into the patriotism demanded of the modern state."[6]

The appeal of all World War I's contending parties to self-defense—what Randolph Bourne termed the "frenzy of self-defense"—raised a number of significant doubts about it.[7] One upshot was that the notion of "self-defense" in international conflicts remained contested after the war. As late as 1928, the United States objected to including the "right of legitimate self-defense" in the Pact of Paris on the grounds that self-defense did not lend itself to precise definition.[8]

Two issues in particular stood out.

One was the question of what *precisely* wars of self-defense were defending. Yes, they were defenses of "sovereignty." But there was complete disagreement as to what the *bearer* of sovereignty was: was it the "state" or the "nation" or the "empire"? The problem was that these were not just different things; they were different *kinds* of things. Defense of the state was defense of a geographical entity ideally defined by borders. Defense of the nation (qua *people*, e.g., "Germans") was defense of a demographic entity, where "borders" makes no sense. Indeed, where "defense of one state" generally meant securing existing borders, "defense of one's nation" often meant disrupting existing borders, to give nations their own state. The turmoil of the interwar years involved the conflict between these principles. Yes, wars were about self-defense, but they were often about *what* merited defending—the state or the nation? Empire only further complicated things (remember that the two World Wars were principally about defenses of empire), beginning with the fact that empires are less a matter of defined borders than vague frontiers, where conflicts have involved where they begin and end.

The other issue was how *expansive* the right to self-defense was. Did the right to self-defense imply the right to the *means* of self-defense? And if so, how far did that extend? On the individual level, this is the question

---

6 Strachan, *Op. cit.*, p. 109.
7 Bourne, Randolph "Below the Battle." *War Is the Health of the State*. Edited by Randolph Bourne, Anecdota Press, 2015.
8 Neff, Stephen C. *War and the Law of Nations: A General History*. Cambridge University Press, 2008, pp. 303–304.

of the relation between the right to self-defense and the right to bear arms. The parallel question for states/nations/empires is the right to armaments, but then the question is whether the right to self-defense implies the right to those resources necessary for such armaments; this was a principal argument of Germany and Japan for their expansive policies in World War II. Japan justified its invasion of Manchuria by appealing to the "rights and interests on which her [Japan's] very existence depends." It was most recently advanced in the United States as an argument for seizing oil resources in the Middle East that were vital to America's national security.

It is worth noting that, posed more abstractly, this was an issue that worried Thomas Hobbes. Hobbes held that the "right to war," by which he meant on a personal level the right to engage in violent self-defense, implied the right to what he called the "helps" of war, by which he meant that everyone needed to exercise that right.

> [B]ecause it is in vain for a man to have a right to the End, if the right to the necessary means be denied him, it follows, that since every Man hath a right to preserve himself, he must also be allowed a right to use all the means, and do all the actions, without which he cannot preserve himself.[9]

Hence, the first law of nature is "That Peace is to be sought after where it may be found; and where not, there to provide our selves for the helps of War." The problem, though, is that what renders the right to self-defense secure creates a general condition of permanent insecurity, or what Hobbes termed a permanent "state of menace"—which he equated with itself a kind of war. Thus, he famously stated, "[A]s long as this natural Right of every man to every thing endureth, there can be no security to any man, (how strong or wise soever he be,) of living out the time, which Nature ordinarily alloweth men to live."[10] Human life is protected in ways that render it less than human, by draining it of meaning—since a state of war has no place for knowledge or culture or industry; the life of man is solitary, poor, nasty, brutish, and short.

Ironically, then, the first great theorist of the right to self-defense insisted that it must be *renounced* in the name of security, or what others have called peace. The "perfect City," he wrote, is one in which "no Citizen hath Right to use his faculties, at his own discretion, for the preservation of

---

9 Hobbes, Thomas. *De Cive (On the Citizen)*. 1642. CreateSpace Independent Publishing, 2015. Chapter 1 (VIII), "Of the State of Men Without Civil Society."
10 Hobbes, Thomas. *Leviathan*. 1651. Penguin Classics, 2017, Chapter XIV, "Of the First and Second Natural Laws, and of Contracts."

himself, or *where the Right of the private Sword is excluded*."[11] Rousseau subsequently posed the obvious question, important to war-skeptics ever since: if civic peace obliges individuals renouncing the right to self-defense as the right to bear arms, why does not international peace oblige nations to renounce the right to self-defense as the right to bear arms? For it is that right as the right to bear arms that compels nations to engage in constant efforts to expand their armaments, in ways that only aggravate the international "state of menace"—leading to the very wars that communities claim to defend themselves against.

The post-World War II settlement did not resolve the issue of state versus nation so much as fudge it—conceptually, via the notion of the "nation-state," and practically, via the enforced migration/expulsions of national groupings to fit their states. For the first time, United Nations Charter established self-defense as a right of nations that licensed armed defense even without the United Nations endorsement. This was endorsed by antiwar communities on the grounds that self-defense, construed in a limited way as *repelling attack*, constituted a significant constraint on how war had been conceived in the past. The problem was that there had always been a more expansive notion of self-defense, construed as *securing survival*, that lent itself to endless inflation.

And this is exactly what happened in the post-World War II era.

Neff has termed this the "self-defense revolution" in international law, wherein self-defense came to "the front and centre of the international stage, as a kind of all-purpose justification for unilateral resorts to armed force." The upshot was to undo much of what the United Nations had sought to achieve, insofar as it was motivated by the desire to seriously constrain if not abolish war through law. As Neff states, the self-defense revolution was a major factor in "undermining the abolition of war."[12] Henceforth, "self-defense" was the dybbuk haunting such attempts to end war.

It did so in part by contributing to the hyperbole—if not hysteria—that came to surround and still surrounds national conflicts, no matter how trivial. Every threat is a threat to national survival and all that the nation "stands for." Not untypical was President Reagan's Secretary of State Alexander Haig's warning that if the United States did not "hold the line" against Nicaragua (a country whose GDP was one-third that of Idaho's), it would mean "the total destruction of our country and everyone in it." My favorite example is America's 1983 invasion of Grenada, "Operation

---

11 Hobbes, *De Cive*, Chapter VI (VIII), "Of the Right of Him, Whether Counsell, or one Man Only, Who Hath the Supreme Power in the City."
12 Neff, *Op. cit.*, p. 315.

Urgent Fury." In it, America unleashed its full military might against a country not quite half the size of Fresno, California, claiming (in Reagan's words) that the "survival" of the "Western hemisphere (sic)" hinged on it. Typical of such actions, most casualties resulted when America accidentally bombed a mental hospital. Seven thousand US soldiers participated in the invasion; over 8,000 medals were awarded for "valor," including 50 to people who never left Washington, DC. And then, of course, we had "Operation Enduring Freedom" and its "preventive" war against Iraq—in the name of "self-defense."

[II]/ War and Personal Self-Defense

The current just war theory discussions of war and self-defense date from David Rodin's book of that name.[13] That book and the discussions it has inspired are too deep and complex for any brief summary. My comments will relate to pacifist concerns.

I take Rodin to have raised two main issues about war as self-defense. One is whether the political right to self-defense can be *modeled* on the personal right to self-defense. The other is whether the political right to self-defense can be *deduced* from the individual right to self-defense. Can war as collective self-defense be understood as the sum of individuals defending themselves? This is of special concern to the pacifist, for, if wars of self-defense are just acts of individual self-defense summed up, then it would seem that the only credible pacifism is one that rejects individual self-defense. This is the argument of Jeff McMahan.[14]

The first question depends entirely on the issues raised above: is the political right to self-defense a right of states, nations, or empires? And how extensive is that right taken to be? Answers to these questions will determine whether, or how much, that right can be modeled on the personal right. There is a certain historical myopia here. Britain's World War II conflict with Germany is taken to be the archetypal just war; but Britain declared war on Germany without having been attacked, as did France—talk of "self-defense" was really about security, identified with certain international arrangements, whose parallel with personal self-defense is tenuous. There is also an asymmetry between the political and the personal that is overlooked, despite its importance to the history of this question. The political right to self-defense, as the right to wage war, clearly presumes the right to the means to do so; war, after all, is *armed* conflict. On the personal level, though, most philosophers are deeply skeptical of any

13 Rodin, David. *War and Self-Defense.* Oxford University Press, 2005.
14 McMahan, Jeff. "Pacifism and Moral Theory." *Diametros*, 23, 2010, pp. 44–68.

essential link between the right to self-defense and the right to bear arms. Discussions elide this issue by ignoring *how* individuals may rightfully kill in self-defense when so entitled.

A further problem is raised by what I have termed the Great Defection.

The right to self-defense is compelling because it is not just the right that one's self be defended but also the right to do the defending *oneself*. How can we ask someone *not* to act in his or her own self-defense when threatened? But we have seen that, while people are adamant about the nation's right to be defended, they have little inclination to do the defending *themselves*. Before America's invasion of Iraq, large majorities supported it as necessary for national self-defense; but when asked their view if they themselves had to fight in it, support for it dropped dramatically. The picture of war as just individual self-defense summed up evokes the picture of a lot of individuals defending themselves. But war is more accurately conceived as *some* people getting *other* people to do the "defending" for them—in conflicts, they would not endorse if they had to do the defending themselves. (All of which raises such questions: if I do not regard a threat as serious enough to defend *myself* against it, at the risk of my life, can I ask someone *else* to defend me against it, at the risk of *their* life?)

Moreover, the picture of war as just a lot of people defending themselves clearly suggests that this is what soldiers are doing. This is assumed in just war theory's recent discussions. In fact, war is the only social practice in which individuals—specifically, soldiers—can apparently be compelled to *renounce* their right to self-defense in the name of some allegedly higher good. Liberal political philosophers have puzzled over this from the beginning. John Locke famously observed how strange it was that a commanding officer "could command a soldier to march up to the mouth of a cannon, or stand in a breach, where he is almost sure to perish" (on the pain of death if the soldier refuses), yet could *not* command "that soldier to give him one penny of his money" or "dispose of one farthing of that soldier's estate, or seize one jot of his goods."[15] A standard response has been that soldiers *consent* to alienate their right to self-defense, but this just raises further questions of a factual sort (Do soldiers "consent" in any meaningful sense, given the draconian consequences of refusing?) and of a normative sort (Is the right to self-defense alienable in this way?). In any event, one reason most people have no interest in serving in the military is that they have no interest in alienating their rights in this way.

---

15 Locke, John. *Second Treatise on Government, 1689*. Henry Regnery, 1955, para. 139. I discuss both Hobbes and Locke on this problem in Ryan, Cheyney "The State and War Making." *For and against the State: New Philosophical Readings*. Edited by John T. Sanders and Jan Narveson, Rowman & Littlefield, 1996.

[III]/ Narveson's Critique

Any discussion of these matters must address Jan Narveson's classic critique of pacifism.[16] I shall be critical of his argument, but let me acknowledge at the start that a very great virtue of his discussion is to raise the question of the language that pacifism employs, specifically whether those pacifists skeptical of self-defense are best understood as employing the language of individual rights. I sketched an alternative view in Chapter 3 and shall conclude with some remarks on Gandhi and Dr. King.

I am afraid that Narveson's discussion is typical of the literature in that he nowhere references any actual pacifists such as Gandhi or Dr. King in his portrayal of "the pacifist" position. His charge against pacifists is a familiar one to philosophers, that their arguments are "incoherent" and pacifists themselves are "deeply confused." His focus is an *inconsistency* he sees in the pacifist position that bears on the question of self-defense.

His argument proceeds thus:

The pacifist's condemnation of killing, he claims, rests on the "right not to be killed" (which is presumably a consequence of the right to life). But inherent in the logic of rights is that "the right not to be subjected to X" implies "the right to defend oneself against X." Hence the right not to be killed implies the right to defend oneself against killing, which Narveson construes as the right to do whatever is necessary to defend oneself—including killing the aggressor. So the pacifist's denial of the right to kill in self-defense conflicts with its grounds for condemning killing in the first place; hence, its inconsistency.

The flaw in this chain of reasoning lies in his claim that the right not to be killed implies the right to do *whatever* is necessary to resist the aggressor up to and including killing him. In this regard, it raises the question of how expansive the right to self-defense is. The right not to be killed implies the right to do *some* things to resist an aggressor. But no pacifist has denied this. On the contrary, both Gandhi and Dr. King endorsed nonviolent resistance. Part of the problem here is that Narveson construes the pacifist as opposing all "force." But pacifists have regarded nonviolence as itself a *type* of force (Gandhi called it "truth force") and have clearly endorsed this type of force as a means to resist being killed.

So the question becomes: does the right not to be killed, as the right to resist lethal aggression, imply the right to resist it by *lethal* force—that is, to *kill* the aggressor, if need be?

---

16 Narveson, Jan. "Pacifism: A Philosophical Analysis." *Today's Moral Problems*. Edited by Richard Wasserstrom, Macmillan Publishing Co., 1975, pp. 450–463.

The pacifist's rejection of the right to resist by lethal force does seem to violate a fairly intuitive principle of proportionality: that in defense of one's rights one may take actions whose severity is equal to the threat against you. Hence, you may kill so as not to be killed. But it is not obvious that resisting X *always* warrants engaging in the *same* X to prevent it. Does resisting nuclear holocaust warrant raining nuclear holocaust on the potential aggressor? For some it does, for others it does not, but the point here is that it is not a matter of the "logic of rights" alone; it involves substantive moral considerations. So too, the pacifist claims, does the question of whether preventing being killed warrants killing oneself.

Moreover, there is a further question of whether Narveson himself properly grasps the "logic of rights" in holding that "the right to X" justifies *whatever* is necessary to defend that right.

Consider the moral stigma of torture. Many construe the right not to be tortured as embodying a deontological prohibition against torturing. And they take this to mean that torturing is impermissible "whatever the consequences," meaning you cannot torture to prevent some larger harm, including more instances of torture; nor can you torture to prevent being tortured yourself. By this logic, then, what the "right not to be subjected to X" means is that one *cannot* do whatever is necessary to prevent it! Indeed, I take it that this is how many just war thinkers regard certain principles of *jus in bello*, like the prohibition on poison gas. Their moral stringency lies in the fact that one cannot violate the principle even to prevent people from violating it against you, that is, attacking you with poison gas.

Again, my point is not that Narveson is wrong to hold that the right not to be killed implies the right to kill to defend that right. My point is that this does not follow from the "logic of rights" per se, hence does not constitute the type of slam dark argument that it claims to be. Pacifist may be wrong, but they are not "incoherent" nor are they themselves "deeply confused."

[IV]/ Gandhi and Dr. King

I want to conclude with some words on Gandhi and Dr. King and self-defense. I have encountered discussions of Dr. King that make a point of insisting that he was *not* a pacifist, presumably to spare him from the stigma that carries. I take this to mean that he did believe one could kill in self-defense. It is fairly clear that Dr. King did not believe in violent self-defense as a political tactic. But what about personal self-defense? The question impresses itself upon us for both Gandhi and Dr. King since I have suggested that they spanned the distinction between personal and political pacifism. I think that any serious attention to their views reveals certain ambiguities. But these ambiguities need not be taken as an uncertainty in their positions. It may reflect the fact that self-defense does not fit into the binary/"Yeah or Nay" logic that talks of the "right" to self-defense and the juridical framework generally require.

The claim that Gandhi "believed in self-defense" appeals to remarks of his like these: "He who cannot protect himself or his nearest and dearest or their honor by non-violently facing death, may and ought to do so by violently dealing with the oppressor." Or, "My non-violence does not admit to running away from danger and leaving dear ones unprotected. Between violence and cowardly flight, I can only prefer violence to cowardice." Neither of these remarks is transparently clear, as sophisticated commentary on them notes. The first says that violence is both permissible and desirable to protect oneself and one's loved ones if someone "cannot" do so by "non-violently facing death." Hence, the nonviolent response remains the morally *preferable* one; plus it need not follow from this that while self-defense is *justified*, it may only be *excused* in light of moral weakness or contingent circumstances. The second remark says that violence is better than cowardly flight, but *non*-cowardly nonviolence is clearly better than both.

I find Dr. King to be even clearer in his reservations about self-defense.

Here is a remark of the same provisional sort as Gandhi's. In "Pilgrimage to Non-Violence," Dr. King writes, "The principle of self-defense, even involving weapons and bloodshed, has never been condemned, even by Gandhi, who sanctioned it for those unable to master pure nonviolence." This is not a claim about what is justified but a claim about what is not to be condemned, and it is as much an argument for why we should master "pure nonviolence" so that we do not have to engage in violent self-defense. But we also know a great deal about Dr. King's practice in this regard. When he was first drawn to nonviolent action in the Montgomery Bus Boycott, he engaged in extensive dialogues with Bayard Rustin about violent self-defense. At the time, he voiced disagreement with Rustin's blanket opposition to violent self-defense. He noted that as a student he had written a paper challenging A.J. Muste's views on nonviolence as incompatible with such self-defense.

By 1964, though, he seems to have come around to Rustin's position. During one campaign, one of the organizers endorsed carrying weapons for personal security saying "we are not totally nonviolent." Writes one authority, "This got back to Martin," who was "totally unhappy because he was totally committed to nonviolence." Later, King tolerated differences about armed self-defense with members of the Student Nonviolent Coordinating Committee (SNCC) and groups like the Louisiana Deacons for Defense injustice. But sometimes he made his disagreements explicit. At the end of a 1966 march, he proclaimed

I'm sick and tired of violence. I'm tired of the war in Vietnam. I'm tired of war and conflict in the world. I'm tired of shooting. I'm tired of hatred. I'm tired of selfishness. I'm tired of evil. I'm not going to use violence, no matter who says it.[17]

The account of this march continues,

> King concluded that violence, "even in self-defense," ultimately created more problems than it solved. The beloved community, "where men can live together without fear," was within reach, but only through "a refusal to hate or kill" in order to "put an end to the chain of violence". The beloved community would require "a qualitative change in our souls" and "a quantitative change in our lives."[18]

Dr. King's pacifist perspective never condemned violent self-defense. On the contrary, responding nonviolently instead of violently was always conceived as a kind of *sacrifice*, which had its impact on others—including the aggressor—from the fact that one was willing to sacrifice. But to conceive of the renunciation of violent self-defense as a "sacrifice" is to assume an entitlement to engage in it of some sort. How, then, should we understand the personal pacifist perspective here?

I construe it as one of *skepticism* toward self-defense. This follows from the sketch of the personal pacifist ethic I provided in Chapter 3, where I suggested that a major challenge for a philosophy of pacifism is to develop this ethic more fully.

---

17 Garrow, David J. "Bearing the Cross: Martin Luther King, Jr., and the Southern Christian Leadership Conference." *Journal of Southern History*, 54(1), 1988, p. 135. https://doi. org/10.2307/2208553.
18 King, Martin Luther. "Nonviolence: The Only Road to Freedom." *Ebony*, October 1966, pp. 27–30. Cited in Strain, Christopher Barry. "Civil Rights and Self-Defense: The Fiction of Nonviolence, 1955–1968." PhD Dissertation, University of California, 2000.

Bibliography

Addams, Jane. *Peace and Bread in Time of War*. University of Illinois Press, 2002.

Akst, Daniel. *War by Other Means: How the Pacifists of World War II Changed America for Good*. Melville House, 2022.

Alexandra, Andrew. "Political Pacifism." *Social Theory and Practice*, vol. 29 no. 4, 2003, pp. 589–606.

Anderson, Benedict. *Imagined Communities: Reflections on the Origin and Spread of Nationalism*. Verso, 2006.

Anderson, Perry. *Passages from Antiquity to Feudalism*. Verso, 1974.

Andrew, Barbara. "The Psychology of Tyranny: Wollstonecraft and Woolf on the Gendered Dimension of War." *Hypatia*, vol. 9 no. 2, 1994, pp. 85–101.

Angell, Norman. *The Great Illusion: A Study of the Relation of Military Power to National Advantage*. 4th ed. CreateSpace Independent Publishing Platform, 2015.

Anscombe, Gertrude Elizabeth Margaret. "War and Murder." *Moral Problems*. Edited by James Rachels. Harper & Row, 1971.

Anzalone, Kyle. "General Milley Predicts Grim Future of Deadly Great Power Wars Fought in Cities." *Antiwar.com*, 22 May 2022, https://news.antiwar.com/2022/05/22/general-milley-predicts-grim-future-of-deadly-great-power-wars-fought-in-cities/.

Appy, Christian G. *Working-Class War: American Combat Soldiers and Vietnam*. University of North Carolina Press, 1993.

Arendt, Hannah. *The Human Condition*. The University of Chicago Press, 1958, pp. 180–181.

Arendt, Hannah. *On Violence*. Houghton Mifflin Harcourt, 1970.

Aron, Raymond. *D'Une Sainte Familla a l'autre* (Cited in Judt, Tony. *The Burden of Responsibility: Blum, Camus, Aron, and the French Twentieth Century*). University of Chicago Press, 2008, p. 13.

Arrighi, Giovanni. "The Three Hegemonies of Historical Capitalism." *Review* (Fernand Braudel Center), vol. 13 no. 3, 1990, pp. 365–408.

Arrighi, Giovanni, and Silver Beverly. *Chaos and Governance in the Modern World System*. University of Minnesota Press, 1999.

Arrighi, Giovanni. *The Long Twentieth Century: Money, Power and the Origins of Our Times*. Verso, 2010.

Ashworth, A. E. *Trench Warfare 1914–1918: The Live and Let Live System*. Palgrave Macmillan, 1980.

Bacevich, Andrew. *American Empire: The Realities and Consequences of U.S. Diplomacy*. Harvard University Press, 2002.

Bagehot, Walter. *Physics and Politics: Or Thoughts on the Application of the Principles of Natural Selection and Inheritance to Political Society*. Legare Street Press, 2021.

Bainton, Roland. *Christian Attitudes Toward War and Peace: A Historical Survey and Critical Re-Evaluation*. Abingdon Press, 1960.

Baker, Nicholson. *Human Smoke: The Beginnings of World War II, the End of Civilization*. Simon and Schuster, 2008.

Baptist, Edward E. *The Half Has Never Been Told: Slavery and the Making of American Capitalism*. Hachette, 2016, pp. 9447–9448.

Bazargan, Saba. "Morally Heterogeneous Wars." *Philosophia*, vol. 41, 2013, pp. 959–975.

Bazargan, Saba. "Varieties of Contingent Pacifism." *How We Fight*. Edited by H. Frowe and G. Lang. Oxford University Press, 2015, pp. 1–17.

Beckert, Sven. *Empire of Cotton: A Global History*. Alfred Knopf, 2014.

Black, Jeremy. *A Military Revolution? Military Change and European Society, 1550–1800*. Red Globe Press, 1991.

Bobbitt, Phillip. *Terror and Consent*. Anchor Books, 2008.

Bobbitt, Philip. "Postscript—The Indian Summer." *The Shield of Achilles*. 2002. Reprint ed., Anchor Books, 2009, pp. 819–824.

Bockman, Chris. "Why the French State Has a Team of UFO Hunters." *BBC News*, November 2014, www.bbc.com/news/magazine-29755919.

Bonney, Richard. *The Rise of the Fiscal State in Europe, 1200–1815*. Oxford University Press, 1999.

Bourne, Randolph. "Below the Battle." *War Is the Health of the State*. Edited by Bourne Randolph. Anecdota Press, 2015.

Bourne, Randolph. *War Is the Health of the State*. 1918. Anecdota Press, 2015.

Boyle, Richard. *GI Revolts: The Breakdown of the U.S. Army in Vietnam*. United Front Press, 1972.

Braker, Regina. "Bertha von Suttner's Spiritual Daughters: The Feminist Pacifism of Anita Augspurg, Lida Gustava Heymann, and Helene Stöcker at the International Congress of Women at the Hague, 1915." *Women's Studies International Forum*, vol. 18 no. 2, 1995, pp. 103–111.

Braudel, Fernand. *Civilization and Capitalism, 15th-18th Century, vol. III*. University of California Press, 1984, pp. 619–633.

Brewer, John. *The Sinews of Power: War, Money and the English State, 1688–1783*. Harvard University Press, 1990.

Brittain, Vera. *Thrice a Stranger*. Victor Gollancz Ltd, 1939.

Brock, Peter. *Pacifism in Europe to 1914*. Princeton University Press, 1972.

Brock, Peter. *Freedom from Violence: Sectarian Nonresistance from the Middle Ages to the Great War*. University of Toronto Press, 1991.

Brock, Peter, and Young Nigel. *Pacifism in the 20th Century*. Syracuse University Press, 1999.

Brown, C. "Reflections on the 'War on Terror', Two Years on." *International Politics*, vol. 41, 2004, pp. 51–64.

Brown, Vincent. *Tacky's Revolt: The Story of an Atlantic Slave War*. Harvard University Press, 2022.

Brueggemann, Walter. *The Prophetic Imagination*. Fortress Press, 2001.

Bruns, Robert. "Blacks, Women Retreat from Army." *Associated Press*, 9 March 2005.

Buber, Martin. *I and Thou*. Translated by Ronald Smith. Charles Scribner's Sons, 1958.

Burbank, Jane, and Cooper Frederick. *Empires in World History: Power and the Politics of Difference*. Princeton University Press, 2011.

Butler, Judith. *The Force of Non-Violence: An Ethico-Political Bind*. Verso, 2021.

Cady, Duane. *From Warism to Pacifism: A Moral Continuum*. 2nd ed. Temple University Press, 2010.

Cavanaugh, William. "A Fire Strong Enough to Consume the House: The Wars of Religion and the Rise of the State." *Modern Theology*, vol. 11 no. 4, 1995, pp. 397–420.

Cave, Damien, Pierson David, and Buckley Chris. "China and U.S. Lay Out Rival Visions for Asia as Ships Nearly Collide." *New York Times*, 4 June 2023.

Cawston, Amanda. " 'Shorn of All Camouflage': Re-assessing the Problem of Violence." Doctoral Thesis, Cambridge University, 2014.

Ceadel, Martin. *Thinking about Peace and War*. Oxford University Press, 1987.

Césaire, Aimé. *Discourse on Colonialism*. Monthly Review Press, 2001.

Chamberlin, Paul Thomas. *The Cold War's Killing Fields: Rethinking the Long Peace*. Harper Collins, 2019.

Chand, Mool. *Nationalism and Internationalism of Gandhi, Nehru, and Tagore*. M.N. Publishers and Distributors, 1989.

Chapman, Gary. "An Introduction to the Revolution in Military Affairs." *Proc. XV Amaldi Conference on Problems in Global Security*, Helsinki, Finland, September 2003.

Chesterman, Simon. *Civilians in War*. Lynne Rienner Publishers, 2001.

Christoyannopoulos, Alexandre. *Tolstoy's Political Thought: Christian Anarcho-Pacifist Iconoclasm Then and Now*. 1st ed. Routledge, 2021.

Christoyannopoulos, Alexandre. "Pacifism and Nonviolence: Discerning the Contours of an Emerging Multidisciplinary Research Agenda." *Journal of Pacifism and Nonviolence*, vol. 1 no. 1, 2023.

Chung, Tan, et al., editors. *Tagore and China*. 1st ed. SAGE Publications Pvt. Ltd, 2011.

Churchill, Ward. *Pacifism as Pathology*. Arbeiter Ring Publishing, 1998.

Clausewitz, Carl. *On War*. Princeton University Press, 1989.

Constant, Benjamin. *Constant: Political Writings*. Translated by Biancmaria Fontana. Cambridge University Press, 1988.

Cortright, David. *Soldiers in Revolt: The American Military Today*. Doubleday, 1975.

Cortright, David. *Peace: A History of Movements and Ideas*. Cambridge University Press, 2008.

Curti, Merle. *Bryan and World Peace*. Garland, 1972.

Darrow, Clarence. *Resist Not Evil*. Ludwig Von Mises Institute, 2007.

Debs, Eugene V. *Canton Speech*. Socialist Party of the United States, 1918, https://college.cengage.com/history/ayers_primary_sources/eugene_canton-speech_1918.htm.

D'Emilio, John. *Lost Prophet: The Life and Times of Bayard Rustin*. 2nd ed. University of Chicago Press, 2004.

De Vattel, Emmerich. *The Law of Nations or the Principles of Natural Law Applied to the Conduct and Affairs of Nations and Sovereigns*. Translated by Joseph Chitty. Lonang Institute, 2003.

De Warren, Nicolas. "The First World War, Philosophy, and Europe." *Tijdschrift voor Filosofie*, 2014, pp. 715–737.

De Warren, Nicolas, and Thomas Vongehr, editors. *Philosophers at the Front. Phenomenology and the First World War*. Leuven University Press, 2018.

Deudney, Daniel H. "Nuclear Weapons and the Waning of the Real-State." *Daedalus*, vol. 124 no. 2, 1995.

Deudney, Daniel H. *Bounding Power: Republican Security Theory from the Polis to the Global Village*. Princeton University Press, 2007.

Dobos, Ned. *Ethics, Security, and The War-Machine: The True Cost of the Military*. Oxford University Press, 2020.

Doshi, Rush. *The Long Game: China's Grand Strategy to Displace American Order*. Kalorama, 2022.

Downing, Brian. *The Military Revolution and Political Change: Origins of Democracy and Autocracy in Early Modern Europe*. Princeton University Press, 1992.

Du Bois, William Edward Burghardt. "The African Roots of War." *The Atlantic*, May 1915, www.theatlantic.com/magazine/archive/1915/05/the-african-roots-of-war/528897/.

Duffeld, Mark. *Global Governance and the New Wars: The Merging of Development and Security*. Zed Books, 2001.

Dutch, Steven. "The Largest Act of Environmental Warfare in History." *Environmental & Engineering Geoscience*, vol. 15 no. 4, 2009, pp. 287–297.

Dworkin, Ronald. *Life's Dominion*. Knopf, 1993.

Dyck, Cornelius. *An Introduction to Mennonite History*. Herald Press, 1967.

Eakin, Emily. "It Takes an Empire." *New York Times*, 2 April 2002, https://wsarch.ucr.edu/wsnmail/2002/msg00510.html.

Edgerton, David. *Warfare State: Britain, 1920–1970*. Cambridge University Press, 2006.

Eghigian, Greg, and Berg Matthew, editors. *Sacrifice and National Belonging in Twentieth-Century Germany*. Texas A&M University Press, 2002.

Elbridge, Colby. *The Strategy of Denial: American Defense in an Age of Great Power Conflict*. Yale University Press, 2021.

Elkins, Caroline. *Legacy of Violence: A History of the British Empire*. Vintage, 2023.

Enloe, Cynthia. *Globalization and Militarism: Feminists Make the Link*. Rowman & Littlefield Publishers, 2016.

Ertman, Thomas. *The Birth of Leviathan*. Cambridge University Press, 1997.

Evans, Peter, et al. editors. *Bringing the State Back*. Cambridge University Press, 1985.

Falk, Richard. "Defining a Just War Ends and Means." *Nation Magazine*, 11 October 2001, https://www.thenation.com/article/archive/defining-just-war/

Falk, Richard, and Kim Samuel S. *The War System: An Interdisciplinary Approach*. Routledge, 2020.

Fanon, Frantz. *The Wretched of the Earth*. Grove Weidenfeld, 1963.

Ferejohn, John, and Rosenbluth Francis McCall. Forged *Through Fire: War, Peace, and the Democratic Bargain*. Liveright, 2016.

Ferguson, Niall. *The Cash Nexus: Economics and Politics from the Age of Warfare through the Age of Welfare, 1700–2000*. Basic Books, 2002.

Ferguson, Niall. *The War of the World: Twentieth Century Conflict and the Descent of the West*. Penguin Press, 2006.

Fiala, Andrew, editor. *Routledge Handbook of Pacifism and Nonviolence*. Routledge, 2018.

Fiala, Andrew. *Transformative Pacifism: Critical Theory and Practice*. Bloomsbury, 2018.

Foucault, Michel. *Society Must Be Defended: Lectures at the College De France, 1975–1976*. Picador, 2003.

French, Howard. *Born in Blackness: Africa, Africans, and the Making of the Modern World, 1471 to the Second World War*. Liveright, 2021.

Fussell, Paul. *Wartime: Understanding and Behavior in the Second World War*. Oxford University Press, 1990.

Gandhi, Mahatma. *The Collected Works of Mahatma Gandhi*, Vol. 10. Publication Division, Government of India, 1965.

Garrison, William Lloyd. *William Lloyd Garrison on Non-Resistance, together with a Personal Sketch by His Daughter Fannie Garrison Villard and a Tribute by Leo Tolstoy*. The Nation's Press Co., 1924.

Garrison, William Lloyd. *William Lloyd Garrison and the Fight against Slavery: Selections from the Liberator*. Edited by Cain William, 1st ed. Bedford/St. Martin's Press, 1994.

Garrow, David J. "Bearing the Cross: Martin Luther King, Jr., and the Southern Christian Leadership Conference." *Journal of Southern History*, vol. 54 no. 1, 1988, p. 135, https://doi.org/10.2307/2208553.

Gay, Peter, *The Enlightenment: The Science of Freedom*. Revised ed., W. W. Norton & Company, 1996.

Gelderloos, Peter. *How Nonviolence Protects the State*. South End Press, 2007.

Gerasimov, Valery. "The Value of Science in Prediction." *Military-Industrial Kurier*, 2013, http://inmoscowsshadows.wordpress.com/2014/07/06/the-Gerasimov-doctrine-and-Russian-non-linear-war/.

Gettleman Marvin, et al. *Vietnam and America: A Documented History*. Grove Press, 1995.

Geyer, Michael. "'There Is a Land Where Everything Is Pure: Its Name Is Land of Death': Some Observations on Catastrophic Nationalism." *Sacrifice and National Belonging in Twentieth-Century Germany*. Edited by Eghigian Greg and Berg Matthew. Texas A&M University Press, 2002, pp. 120–141.

Gilbert, Martin. *A History of the Twentieth Century 1900–1933*. Vol. 1, William Morrow and Company, 1997.

Gittings, John. "The Growth of Peace Consciousness: From the Enlightenment to The Hague." *The Glorious Art of Peace: From the Iliad to Iraq*. Oxford University Press, 2012.

Glover, Jonathan. *Humanity: A Moral History of the 20th Century*. Yale University Press, 2001.

Goldstein, Joshua. *War and Gender: How Gender Shapes the War System and Vice Versa*. Cambridge University Press, 2009.

Graeber, David, and Wengrow David. *The Dawn of Everything: A New History of Humanity*. Farrar, Straus and Giroux, 2021.

Graham, Bradley, and White Josh. "Abizaid Credited With Popularizing the Term 'Long War'." *The Washington Post*, February 3, 2006. https://www.washingtonpost.com/archive/politics/2006/02/03/abizaid-credited-with-popularizing-the-term-long-war/7d26310e-31e3-4be6-a910-7d898a645e73/.

Gray, Chris Hables. *Post-Modern War: The New Politics of Conflict*. Routledge, 2007.

Gray, Jesse Glen. *The Warriors: Reflections on Men in Battle*. Bison Books, 1998.

Green, Thomas Hill. *Lectures on the Principles of Political Obligation (1883)*. Batoche Books, 1999.

Hale, John R. *Lords of the Sea: The Epic Story of the Athenian Navy and the Birth of Democracy*. Penguin, 2009.

Harvey, David. *The New Imperialism*. Oxford University Press, 2003.

Hastings, Max. *All Hell Loose: The World at War 1939–1945*. Harper, 2011.

Hauerwas, Stanley. *The Peaceable Kingdom: A Primer in Christian Ethics*. University of Notre Dame Press, 1991.

Hauerwas, Stanley. "Sacrificing the Sacrifices of War." *Journal of Religion, Conflict, and Peace*, vol. 1 no. 1, 2007, www.religionconflictpeace.org/volume-1-issue-1-fall-2007/sacrificing-sacrifices-war.

Hauerwas, Stanley. *War and the American Difference: Theological Reflections on Violence and National Identity*. Baker Academic, 2011.

Hauerwas, Stanley, et al. "The Case for Abolition of War in the Twenty-First Century." *Journal of the Society of Christian Ethics*, vol. 25 no. 2, 2005, pp. 17–35.

Hedges, Chris. *War Is a Force That Gives Us Meaning*. Public Affairs, 2002.

Hegel, G.W. *Hegel's Philosophy of Right. 1820*. Translated by T M Knox, Clarendon Press, 1952, paragraph 93.

Heinl, Jr., Colonel Robert D. "The Collapse of the Armed Forces." *Armed Forces Journal*, 7 June 1971, pp. 30–37. https://msuweb.montclair.edu/~furrg/Vietnam/heinl.html.

Hinsley, F.H. *Nationalism and the International System*. Hodder and Stoughton, 1973.

Hobbes, Thomas. *De Cive (On the Citizen). 1642*. CreateSpace Independent Publishing, 2015.

Hobbes, Thomas. *Leviathan. 1651*. Penguin Classics, 2017.

Hobsbawm, Eric. *Age of Extremes: The Short Twentieth Century, 1914–1991*. Abacus, 1995.

Hobsbawm, Eric. *Nations and Nationalism since 1780*. Cambridge University Press, 2012.

Hoffman, David. *The Dead Hand: The Untold Story of the Cold War Arms Race and Its Dangerous Legacy.* 1st ed. Anchor, 2010.

Hoffman, Frank. *Conflict in the 21st century: The Rise of Hybrid Wars.* Potomac Institute for Policy Studies, 2007.

Holmes, Robert. *Pacifism: A Philosophy of Nonviolence.* Bloomsbury Academic, 2017.

Howard, Michael. "War and the Nation–State." *Daedalus*, vol. 108 no. 4, 1979, pp. 101–110.

Howard, Michael. *The Lessons of History.* Yale University Press, 1991.

Howard, Michael. "When Are Wars Decisive?" *Survival*, vol. 41 no. 1, 1999, p. 129.

Howard, Michael. "What's in a Name? How to Fight Terrorism." *Foreign Affairs*, vol. 81 no. 1, 2002, pp. 8–13.

Howard, Michael. *War in European History.* Oxford University Press, 2009.

Howes, Dustin Ells. "The Failure of Pacifism and the Success of Nonviolence." *Perspectives on Politics*, vol. 11 no. 2, 2013, pp. 427–446.

Immerwahr, Daniel. *How to Hide an Empire: A History of the Greater United States.* Picador, 2020.

Jackson, Richard. "Bringing Pacifism Back Into International Relations." *Social Alternatives*, vol. 33 no. 4, 2014, pp. 63–66.

Jackson, Richard. "Pacifism: The Anatomy of a Subjugated Knowledge." *Critical Studies on Security*, vol. 6 no. 2, 2018, pp. 160–175.

Jackson, Sarah J. "Black Lives Matter and the Revitalization of Collective Visionary Leadership." *Leadership*, vol. 17 no. 1, 2021, pp. 8–17.

James, Harold. *The Roman Predicament: How the Rules of International Order Create the Politics of Empire.* Princeton University Press, 2006.

James, William. "Remarks at the Peace Banquet." *Memories and Studies.* Longman Green and Co., 1911, pp. 299–306.

James, William. *The Moral Equivalent of War.* Obscure Press, 2013.

James, William. *Varieties of Religious Experience, a Study in Human Nature. 1902.* Cross Reach Publications, 2017.

Joas, Hans. *War and Modernity: Studies in the History of Violence in the 20th Century.* Translated by Rodney Livingstone, Polity, 2003.

Johansen, Robert C. "Radical Islam and Nonviolence: A Case Study of Religious Empowerment and Constraint among Pashtuns." *Journal of Peace Research*, vol. 34 no. 1, 1997, pp. 53–71.

Johnson, Chalmers. *Blowback: The Cost and Consequences of American Empire.* Metropolitan Books, 2000.

Johnson, Chalmers. *The Sorrows of Empire: Militarism, Secrecy, and the End of the Republic.* Metropolitan Books, 2004.

Johnson, Chalmers. *Nemesis: The Last Days of the American Republic.* Metropolitan Books, 2006.

Johnson, James Turner. "Rationalizing the Hell of War." *Worldview*, vol. 17, 1974, pp. 43–47.

Joy, Bill. "Why the Future Doesn't Need Us: Our Most Powerful 21st Century Technologies—Robotics, Genetic Engineering, and Nanotech—Are Threatening to Make Humans an Endangered Species." *Wired*, vol. 8 no. 4, 2000, www.wired.com/2000/04/joy-2/.

Judt, Tony. *Postwar: A History of Europe since 1945*. Penguin Press, 2005.

Judt, Tony. *The Burden of Responsibility: Blum, Camus, Aron, and the French Twentieth Century*. University of Chicago Press, 2008.

Kaldor, Mary. "Protective Security or Protection Rackets? War and Sovereignty." *Arguments for a Better World: Essays in Honor of Amartya Sen, Volume 2: Society, Institutions, and Development*. Edited by Basu Kaushik and Kanbur Ravi. Oxford University Press, 2008, pp. 470–487.

Kaldor, Mary. "Inconclusive Wars: Is Clausewitz Still Relevant in These Global Times?" *Global Policy*, vol. 1 no. 3, 2010, pp. 271–281.

Kaldor, Mary. *New and Old Wars: Organized Violence in a Global Era*. 3rd ed. Stanford University Press, 2012.

Kant, Immanuel. *Perpetual Peace and other Essays on Politics, History, and Morals*. Translated by Ted Humphrey. Hackett Publishing Co., 1983.

Kapoor, Sudarshan. 1992. *Raising Up a Prophet: The African-American Encounter with Gandhi*. Beacon Press, 2012.

Kazin, Michael. *A Godly Hero: The Life of William Jennings Bryan*. Random House, 2006.

Keegan, John. *A History of Warfare*. Alfred A. Knopf, 1997.

Keen, David. *Hidden Functions of the War on Terror*. Pluto Press, 2007.

Keen, David. *Useful Enemies: When Waging Wars Is More Important Than Winning Them*. Manas Publications, 2014.

Kennedy, Paul. *The Rise and Fall of the Great Powers*. Knopf Doubleday, 1987.

King, Martin Luther. *Stride Toward Freedom*. Ballantine, 1958.

King, Martin Luther. "Nonviolence: The Only Road to Freedom." *Ebony*, October 1966, pp. 27–30.

Koestler, Arthur. *Darkness at Noon*. Penguin, 1987.

Kolko, Gabriel. *Century of War: Politics, Conflicts, and Society since 1914*. New Press, 1994.

Kolko, Gabriel. *Another Century of War?* New Press, 2004.

Kraditor, Aileen S. *Means and Ends in American Abolitionism: Garrison and His Critics on Strategy and Tactics 1834–1850*. Ivan R. Dee, 1989.

Kundrus, Birthe. "Die Kolonien—'Kinder des Gefühls und der Phantasie'." *Phantasiereiche*. Edited by idem, 2021, pp. 7–18 [Cited in Overy, Richard J. *Blood and Ruins*. Allen Lane, an imprint of Penguin Books].

Kundu, Kalyan. "Rabindranath Tagore and World Peace." *Asiatic*, vol. 4 no. 1, 2010, p. 81.

LaFeber, Walter. "An End to Which Cold War." *The End of the Cold War: Its Meaning and Implications*. Edited by Michael J. Hogan. Cambridge University Press, 1992.

Lary, Diana. "Drowned Earth: The Strategic Breaching of the Yellow River Dyke, 1938." *War in History*, vol. 8 no. 2, 2001, pp. 191–207.

Lazar, Seth. "War." Edited by Edward N. Zalta. Stanford Encyclopedia of Philosophy, 2016, https://plato.stanford.edu/entries/war.

Lazar, Seth. "Just War Theory: Revisionists Vs Traditionalists." *Annual Review of Political Science*, vol. 20, 2017, pp. 37–54.

Levinas, Emmanuel. *Totality and Infinity*. Translated by A. Lingis. Duquesne University Press, 1969.

Liebknecht, Karl. *Militarism and Anti-Militarism: With Special Regard to the International Young Socialist Movement*. Translated by Grahame Lock, Rivers Press Limited, 1973.

Linebaugh, Peter. and Rediker Marcus. *The Many-Headed Hydra: Sailors, Slaves, Commoners, and the Hidden History of the Revolutionary Atlantic*. Beacon Press, 2013.

Lippman, Walter. *The Public Philosophy. 1955*. Routledge, 2017.

Locke, John. *Second Treatise on Government. 1689*. Chicago: Henry Regnery, 1955.

Luard, Evan. 1986. *War in International Society*. Yale University Press, 1986.

Lutz, Catherine. "Making War at Home in the United States: Militarization and the Current Crisis." *American Anthropologist*, vol. 104 no. 3, pp. 723–735.

Luxemburg, Rosa. "The Meaning of Pacifism." *Socialist Appeal*, vol. II no. 14, 1938. www.marxists.org/history/etol/newspape/themilitant/socialist-appeal-1938/v02n14/luxemburg.htm.

Mack, Eric. "Astronomers Just Updated the Odds a Massive Asteroid Will Hit Earth in The Future." *Forbes*, 29 March 2022, www.forbes.com/sites/ericmack/2022/03/29/astronomers-just-updated-the-odds-a-massive-asteroid-will-hit-earth-in-the-future/?sh=538be2ec3a9c.

Malesevic, Sinisa. *The Sociology of War and Violence*. Cambridge University Press, 2010.

Manicas, Peter T. *War and Democracy*. Blackwell Publishing, 1989.

Mann, Michael. "War and Social Theory." *The Sociology of War and Peace*. Edited by Creighton Colin and Shaw Martin. McMillan, 1987.

Mann, Michael. *On Wars*. Yale University Press, 2003.

Mann, Michael. *The Sources of Social Power: Volume 2, The Rise of Classes and Nation-States, 1760–1914*. Cambridge University Press, 2012.

Mann, Michael. *The Sources of Social Power: Volume 3, Global Empires and Revolution, 1890–1945*. Cambridge University Press, 2012.

Marcel, Gabriel. *Homo Viator: Introduction to a Metaphysic of Hope*. Henry Regency, 1951.

Marcel, Gabriel. "Autobiographical Essay." *The Philosophy of Gabriel Marcel: The Library of Living Philosophers, 17*. Edited by Paul Arthur Schilpp and Lewis Edwin Hahn. Open Court, 1984.

Martin, Adrienne, editor. *The Routledge Handbook of Love in Philosophy*. Routledge, 2019.

Marvin, Carolyn, and Ingle David. *Blood Sacrifice and the Nation: Totem Rituals and the American Flag*. Cambridge University Press, 1999.

May, Larry. *Contingent Pacifism: Revisiting Just War Theory*. Cambridge University Press, 2015.

Mayer, Henry. *All on Fire: William Lloyd Garrison and the Abolition of Slavery*. 1st ed. St Martin's Press, 1998.

Mayerfeld, Jamie. "In Defense of the Absolute Prohibition of Torture." *Public Affairs Quarterly*, vol. 22 no. 2, 2008, pp. 109–128.

Mazower, Mark. *Dark Continent: Europe's Twentieth Century*. Vintage, 2000.

McAndrew, Tara. "Local Icon Shifts To 'The Most Dangerous Woman in America.'" *Illinois Public Media*, 11 July 2017, https://will.illinois.edu/news/story/illinois-issues-local-icon-shifts-to-the-most-dangerous-woman-in-america.

McCoy, Alfred. *To Govern the Globe: World Orders and Catastrophic Change.* Haymarket Books, 2021.

McMahan, Jeff. *Killing in War.* Oxford University Press, 2009.

McMahan, Jeff. "Pacifism and Moral Theory." *Diametros*, no. 23, 2010, pp. 44–68.

McNeill, William H. *The Rise of the West: A History of the Human Community.* University of Chicago Press, 1963.

McNeill, William H. *Keeping Together in Time: Dance and Drill in Human History.* Harvard University Press, 1995.

Mills, Charles. Wright. *The Causes of World War Three.* Greenwood Press, 1976.

Milton, John, "Sonnet 15: Fairfax." *The Sonnets of Joh Milton.* Legare Street Press, 2022.

Mishra, Pankaj. *The Ruins of Empire: The Revolt against the West and the Remaking of Asia.* Reprint ed., Penguin Books Ltd, 2013.

Moyn, Samuel. *Humane: How the United States Abandoned Peace and Reinvented War.* Farrar, Straus and Giroux, 2021.

Mueller, John. "The Obsolescence of Major War Bulletin of Peace Proposals." *Bulletin of Peace Proposals*, vol. 21 no. 3, 1990, pp. 321–328.

Mueller, John. "War Has Almost Ceased to Exist: An Assessment." *Political Science Quarterly*, vol. 124 no. 2, 2009, pp. 297–321.

Mukerjee, Madhusree. *Churchill's Secret War: The British Empire and the Ravaging of India During World War II.* Penguin Books, 2018.

Mumford, Lewis. "Gentlemen: You Are Mad!" *Saturday Review of Literature*, 2 March 1946, pp. 5–6.

Munkler, Herfried. *The New Wars.* Polity Press, 2005.

Murphy, Paul L. *World War I and the Origin of Civil Liberties in the United States.* Norton, 1979.

Muthu, Sankar. *Enlightenment against Empire.* Princeton University Press, 2003.

Nagel, Thomas. "War and Massacre." *Philosophy and Public Affairs*, vol. 1, 1972, pp. 123–144.

Narveson, Jan. "Pacifism: A Philosophical Analysis." *Today's Moral Problems.* Edited by Wasserstrom Richard. Macmillan Publishing Co., 1975, pp. 450–63.

Neff, Stephen C. *War and the Law of Nations: A General History.* Cambridge University Press, 2008.

Niebuhr, H. Richard. *The Meaning of Revelation.* Macmillan, 1941.

Nietzsche, Friedrich. *Human All-Too-Human A Book For Free Spirits, Part II.* Translated by Cohn, Paul V. Part of Nietzsche, Friedrich, *The Complete Works of Friedrich Nietzsche.* Edited by Oscar Levy. The MacMillan Company, 1911.

Norman, Richard. *Ethics, Killing and War.* Cambridge University Press, 1995.

Orosco, José-Antonio. "Pacifism as Pathology." *The Routledge Handbook of Pacifism and Nonviolence.* Edited by Andrew Fiala. 1st ed. Routledge, 2020.

Orwell, George. *1984.* Secker & Warburg, 1949.

Overy, Richard. *Blood and Ruins.* Allen Lane, an imprint of Penguin Books, 2021.

Paine, Thomas. *The American Crisis*. *1776*. CreateSpace Independent Publishing Platform, 2017.

Parker, Geoffrey. "The Military Revolution, 1560–1660—A Myth?" *Essays in Swedish History*. Edited by Roberts Michael. J. Wyatt Books, 1967.

Parker, Patricia S. *Ella Baker's Catalytic Leadership: A Primer on Community Engagement and Communication for Social Justice*. 1st ed. University of California Press, 2020.

Parkin, Nicholas. "Conditional and Contingent Pacifism: The Main Battlegrounds." *Critical Studies on Security*, vol. 6 no. 2, pp. 193–206.

Parsons, Graham, et al., editors. *To End a War: Essays on Justice, Peace, and Repair*. Cambridge University Press, 2022.

Piketty, Thomas. *Capital in the Twenty-First Century*. Translated by Arthur Goldhammer. Harvard University Press. 2014.

Pitts, Jennifer. *A Turn to Empire: The Rise of Imperial Liberalism in Britain and France*. Princeton University Press, 2005.

Platt, Stephen R. *Autumn in the Heavenly Kingdom: China, the West, and the Epic Story of the Taiping Civil War*. Atlantic Books, 2014.

Podhoretz, Norman. *World War IV: The Long Struggle against Islamofascism*. Doubleday, 2007.

Popper, Karl. *The Open Society and Its Enemies*. Princeton University Press, 2020.

Porter, Bruce. *War and The Rise of The State*. Free Press, 2002.

Ramsey, Paul. "Is Vietnam a Just War?" *Dialog*, 1967, pp. 19–29.

Ramsey, Paul. *Just War: Force and Political Responsibility*. Rowman and Littlefield, 2002.

Ransby, Barbara. *Ella Baker and the Black Freedom Movement: A Radical Democratic Vision*. The University of North Carolina Press, 2003.

Reardon, Betty. *Sexism and the War System*. Syracuse University Press, 1996.

Rediker, Marcus. *Between the Devil and the Deep Blue Sea: Merchant Seamen, Pirates and the Anglo-American Maritime World, 1700–1750*. Cambridge University Press, 1989.

Rediker, M. *Outlaws of the Atlantic: Sailors, Pirates, and Motley Crews in the Age of Sail*. Beacon Press, 2015.

Reese, David "Regulation of Bodies as Gendered Nationalistic Ideology: Physically Wounded Veterans as Political Props." Masters Thesis, University of Oregon, 2015.

Remarque, Erich Maria. *All Quiet on the Western Front*. *1928*. Ballantine Books, 1987.

Rengger, Nicholas. "On the Just War Tradition in the Twenty-First Century." *International Affairs*, vol. 78 no. 2, 2002, pp. 353–363, p. 354.

Rhodes, Richard. *Arsenals of Folly: The Making of the Nuclear Arms Race*. Deckle Edge, 2007.

Roberts, Michael. *Essays in Swedish History*. J Wyatt Books, 1967.

Roberts, Sam. "Kaname Harada, Pearl Harbor Fighter Pilot and, Later, Remorseful Pacifist, Dies at 99." *The New York Times*, 5 May 2016, www.nytimes.com/2016/05/06/world/asia/kaname-harada-pearl-harbor-fighter-pilot-who-became-pacifist-dies-at-99.html.

Robillard, Michael, and Strawser Bradley. *Outsourcing Duty: The Moral Exploitation of the American Soldier*. Oxford University Press, 2022.

Rodin, David. *War and Self-Defense*. Oxford University Press, 2005.

Rovelli, Carlo, and Smerlak Matteo. "A Small Cut in World Military Spending Could Help Fund Climate, Health and Poverty Solutions." *Scientific American*, 17 March 2022, www.scientificamerican.com/article/a-small-cut-in-world-military-spending-could-help-fund-climate-health-and-poverty-solutions/.

Russell, Bertrand. "The Ethics of War." *The International Journal of Ethics*, vol. 25 no. 2, 1915, pp. 127–142.

Rustin, Bayard. "The Negro and Nonviolence." *PA History*, 1942, https://explore-pahistory.com/odocument.php?docId=1-4-184.

Rustin, Bayard. *Time on Two Crosses: The Collected Writings of Bayard Rustin*. Edited by Devon W. Carbado and Don Weise, 2nd ed. Cleis Press, 2015.

Ryan, Cheyney. "Pacifism, Self Defense, and the Possibility of Killing." *Ethics*, vol. 93 no. 3 (Apr., 1983), pp. 508–524.

Ryan, Cheyney. "The State and War Making." *For and Against the State: New Philosophical Readings*. Edited by Sanders John T. and Narveson Jan. Rowman & Littlefield, 1996.

Ryan, Cheyney. *The Chickenhawk Syndrome: War, Sacrifice, and Personal Responsibility*. Roman and Littlefield, 2009.

Ryan, Cheyney. "The One Who Burns Herself for Peace." *Hypatia a Journal of Feminist Philosophy*, vol. 9 no. 2, 2009, pp. 21–39.

Ryan, Cheyney. "Pacifism." *Oxford Handbook on the Ethics of War*. Edited by Lazar Seth and Frowe Helen. Oxford University Press, 2016.

Ryan, Cheyney. "Bearers of Hope: On the Paradox of Non-Violent Action." *The Ethics of Soft War*. Edited by Michael Gross and Tami Meisels. Cambridge University Press, 2017.

Ryan, Cheyney. "The Pacifist Critique of Just War Theory." *The Routledge Handbook of Pacifism and Nonviolence*. Edited by Fiala Andrew. Routledge, 2017.

Ryan, Cheyney. "The Morning Stars Will Sing Together: Compassion, Nonviolence, and the Revolution of the Heart." *The Routledge Handbook of Love in Philosophy*. Edited by Martin Adrienne. Routledge, 2019.

Ryan, Cheyney. "War, Hostilities, Terrorism: A Pacifist Perspective." *Pacifism's Appeal Ethos, History, Politics*. Edited by Kustermans J., Sauer T., Lootens D. and Segaert B. Palgrave MacMillan, 2019.

Ryan, Cheyney. "'Wretched Nurseries of Unceasing Discord': War, Nationalism, and the Politics of Peace." *Theoretical Inquiries in Law*, vol. 21 no. 2, 2020, pp. 207–228.

Ryan, Cheyney. "The Lament of the Demobilized." *To End a War: Essays on Justice, Peace, and Repair*. Edited by Graham Parsons and Mark Wilson. Cambridge University Press, 2022.

Sahibzada, Imitiaz Ahmad. *The Frontier Gandhi: My Life and Struggle: The Autobiography of Abdul Ghaffar Khan*. Roli Books, 2021.

Samet, Elizabeth D. *Looking for the Good War: American Amnesia and the Violent Pursuit of Happiness*. Farrar, Straus and Giroux, 2021.

Sanders, John T., and Narveson Jan, editors. *For and against the State: New Philosophical Readings*. Rowman & Littlefield, 1996.

Savage, Charlie. "U.S. Discloses Decades of Justice Dept. Memos on Presidential War Power." *New York Times*, 16 September 2022, www.nytimes.com/2022/09/16/us/politics/war-powers-justice-dept-president.html.

Scarry, Elaine. *Thermonuclear Monarchy: Choosing Between Democracy and Doom.* Summary ed., W. W. Norton & Company, 2016.

Scheidel, Walter. *The Great Leveler: Violence and the History of Inequality from the Stone Age to the Twenty-First Century.* Princeton University Press, 2018.

Schell, Jonathan. *The Unconquerable World: Power, Nonviolence, and the Will of the People.* Penguin, 2005.

Schildgen, Robert. *Toyohiko Kagawa: An Apostle of Love and Social Justice.* 1st ed. Centenary Books, 1988.

Schumpeter, Joseph. *Imperialism and Social Classes. 1919.* Ludwig von Mises Institute, 2007.

Sharify-Funk, Meena. "Toward a Global Understanding of Pacifism: Hindu, Islamic, and Buddhist Contributions." *Pacifism's Appeal: Ethos, History, Politics.* Edited by Jorg Kuystermans et al., 1st ed. Palgrave Macmillan, 2019.

Shaw, Martin. *Dialectics of War: An Essay in the Social Theory of Total War and Peace.* Pluto Press, 1988.

Shaw, Martin. "War and the Nation State in Social Theory." *Social theory of Modern Societies: Anthony Giddens and His Critics.* Edited by Held David and Thompson John. Cambridge University Press, 1989.

Shaw, Martin. *Global Society and International Relations: Sociological Concepts and Political Perspectives.* Polity Press, 1994.

Shaw, Martin. *War and Genocide: Organised Killing in Modern Society.* Polity, 2003.

Sheehan, James L. *Where Have All the Soldiers Gone?: The Transformation of Modern Europe.* Houghton Mifflin, 2008.

Sherry, Michael. *The Rise of American Air Power: The Creation of Armageddon.* Yale University Press. 1989.

Sherry, Michael. *In the Shadow of War: The United States since the 1930s.* Yale University Press, 1995.

Shue, Henry. *Basic Rights: Subsistence, Affluence and U.S. Foreign Policy.* Princeton University Press, 1996.

Simon, Scott. "Even Pacifists Must Support This War." *Wall Street Journal,* 2001.

Singh, Nikhil Pal. *Race and America's Long War.* University of California Press, 2019.

Sjoberg, Laura, and Peet Jessica. "A(nother) Dark Side of the Protection Racket." *International Feminist Journal of Politics,* vol. 13 no. 2, 2011, pp. 163–182.

Sjoberg, Laura. "Witnessing the Protection Racket: Rethinking Justice in/of Wars Through Gender Lenses." *International Politics,* vol. 53 no. 3, 2016, pp. 361–384.

Slate, Nico. *Colored Cosmopolitanism: The Shared Struggle for Freedom in the United States and India.* Harvard University Press, 2017.

Smith, Rupert. *The Utility of Force: The Art of War in the Modern World.* Alfred A. Knopf, 2005.

Soelle, Dorothee. *Suffering.* Translated by E. R. Kalin. Fortress Press, 1975.

Soelle, Dorothee. *Revolutionary Patience*. Wipf and Stock, 2003.

Spinelli, Altiero. "For a Free and United Europe. A Draft Manifesto"; Otherwise Titled, "The Manifesto of Ventotene." *English Translation, Union of European Federalists*, 1941, www.federalists.eu/uef/library/books/the-ventotene-manifesto/.

Stoker, Donald. Why *America Loses Wars: Limited War and US Strategy from the Korean War to the Present*. Cambridge University Press, 2022.

Strachan, Hew, et al., editors. *Clausewitz and the Twenty First Century*. 1st ed. Oxford University Press, 2007.

Sumner, Charles. *An Address Before the American Peace Society*. American Peace Society, 1854.

Tagore, Rabindranath. *Nationalism*. MacMillan and Co. Ltd, 1918.

Tarrow, Sidney. *War, States, and Contention: A Comparative Historical Study*. Cornell University Press, 2015.

Teichman, Jenny. *Pacifism and the Just War*. Basil Blackwell, 1986.

Tertullian, Quintus. De Idololatria (On Idolatry). Translated by Rev. S Thelwall, 2015, Chapter XIX, www.documentacatholicaomnia. eu/03d/0160-0220,_Tertullianus,_De_Idolatria_[Schaff],_EN.pdf.

Thurman, Howard. *With Head and Heart: The Autobiography of Howard Thurman*. New Mariner Books, 1981.

Tierney, Dominic *The Right Way to Lose a War: America in an Age of Unwinnable Conflicts*. Little Brown and Company, 2015.

Tilly, Charles. "Where Do Rights Come From?" *Democracy, Revolution, and History*. Edited by Theda Skocpol. Cornell University Press, 2000.

Tilly, Charles. *Coercion, Capital, and European States: AD 990–1992*. Blackwell, 2015.

Tolstoy, Leo. "Letter on the Peace Conference." *Tolstoy's Writings on Civil Disobedience and Nonviolence*. New American Library, 1968, pp. 113–119.

Tolstoy, Leo. *Anna Karenina*. Translated by Rosemary Edmunds. Penguin, 1978.

Tolstoy, Leo. *The Kingdom of God Is Within You*. Translated by Constance Garnett, Independently Published, 2021.

Tuck, Richard. *Natural Rights Theories: Their Origin and Development*. Cambridge University Press, 1982.

Ullman, Harlan. *Anatomy of Failure: Why America Loses Every War It Starts*. Naval Institute Press, 2017.

Ulrich, Beck U. *Risk Society: Towards a New Modernity*. Sage, 1992.

Unger, Mangabeira Roberto. *Passion: An Essay on Personality*. Simon and Schuster, 1986.

Unger, Peter. "I Do Not Exist." *Perception and Identity*, Edited by G. F. MacDonald. Macmillan, 1979, pp. 235–251.

Van Creveld, Martin. *The Rise and Decline of the State*. Cambridge University Press, 1999.

Vartabedian, Ralph. "A Poisonous Cold War Legacy that Defies a Solution." *The New York Times*, 31 May 2023, www.nytimes.com/2023/05/31/us/nuclear-waste-cleanup.html.

Vayrynen, Raimo, editor. *The Waning of Major War: Theories and Debates*. Routledge, 2005.

Vine, David. *Base Nation: How U.S. Military Bases Abroad Harm America and the World*. Metropolitan Books, 2015.

Vine, David. *The United States of War: A Global History of America's Endless Conflicts, from Columbus to the Islamic State*. University of California Press, 2020.

Von Suttner, Bertha. "The Evolution of the Peace Movement." *Nobel Lecture*, 18 April 1906, www.nobelprize.org/prizes/peace/1905/suttner/lecture/.

Von Suttner, Bertha. "International Peace Through the Voice of Women." *Archives of Women's Communication*. Iowa State University, 1912, https://awpc.cattcenter.iastate.edu/2018/10/18/international-peace-through-the-voice-of-women-july-2–1912/.

Waldron, Jeremy. "Torture and Positive Law: Jurisprudence for the White House." *Colombia Law Review*, vol. 105 no. 6, 2005, pp. 1681–1750.

Waldron, Jeremy. *Torture, Terror, and Trade-Offs: Philosophy for the White House*. Oxford University Press, 2010.

Wallace, Rick. "WWII Japanese Pilot's Quest for Redemption." *The Australian*, 3 April 2015, https://www.theaustralian.com.au/subscribe/news/1/?sourceCode=TAWEB_WRE170_a_GGL&dest=https%3A%2F%2Fwww.theaustralian.com.au%2Fnews%2Fworld%2Fwwii-japanese-pilots-quest-for-redemption%2Fnews-story%2Fd5bb377a3d700eabadbbac25993fd578&memtype=anonymous&mode=premium&v21=GROUPA-Segment-2-NOSCORE&V21spcbehaviour=append.

Wallerstein, Immanuel. "The Three Instances of Hegemony in the History of the Capitalist World-Economy." *International Journal of Comparative Sociology*, vol. 24, 1983.

Walsh, David. "Opposition to Iraq War Hitting US Military Recruitment." *World Socialist Web*, March 2005, www.wsws.org/en/articles/2005/03/mili-m12.html.

Walzer, Michael. *Just and Unjust Wars*. Basic Books, 1977.

Walzer, Michael. "The Triumph of Just War Theory (and the Dangers of Success)." *Social Research*, vol. 69 no. 4, 2002, pp. 925–944.

Wasserstrom, Richard, editor. *Today's Moral Problems*. Macmillan Publishing Co., 1975.

Watson Institute Costs of War Project. "How Death Outlives War: The Reverberating Impact of the Post-9/11 Wars on Human Health." *Report*, May 2023, https://watson.brown.edu/costsofwar/files/cow/imce/papers/2023/Indirect%20Deaths.pdf.

Weinberg, Gerhard. *A World at Arms—A Global History of World War II*. Cambridge University Press, 1994.

Wells, Tom. *The War Within: America's Battle Over Vietnam*. Henry Holt, 1994.

West, Brad West, and Mattherman Steve. "Towards a Strong Program in the Sociology of War, the Military and Civil Society." *Journal of Sociology*, vol. 52 no. 3, 2016, pp. 482–499.

Whitlock, Craig. "At War with the Truth." *Washington Post*, 9 December 2019.

Whitman, Walt. "Death of Abraham Lincoln Lecture." *Complete Works of Walt Whitman*. Delphi Classics, 2012, p. 1949.

Whitney, Craig R. "U.S. Protests Invective by Saigon on McGovern." *The New York Times*, 11 September 1972, www.nytimes.com/1972/09/11/archives/us-protests-invective-by-saigon-on-mcgovern.html.

Wills, Gary. *Bomb Power: The Modern Presidency and the National Security State*. Penguin, 2011.

Wimmer, Andreas. *Waves of War: Nationalism, State Formation, and Ethnic Exclusion in the Modern World*. Cambridge University Press, 2013.

Wink, Walter. *The Powers That Be*. Harmony, 2010.

Wood, Ellen Meiksins. *Liberty and Property: A Social History of Western Political Thought from Renaissance to Enlightenment*. Verso, 2012.

Woolman, John. *The Journal of John Woolman and A Plea for the Poor*. Wipf and Stock, 1998.

Woolsey, Robert James. *WWIV: Who We're Fighting- And Why: Vol. 4*. Rich. J. Global L. & Bus. 1, 2004.

Xu, Guoqi. *Asia and the Great War: A Shared History*. 1st ed. Oxford University Press, 2017.

Yoder, John Howard. *Preface to Theology: Christology and Theological Method*. Baker Publishing Group, 2002.

Yoder, John Howard. *The War of the Lamb: The Ethics of Nonviolence and Peacemaking*. Brazos Press, 2009.

Yoder, John Howard. *Nonviolence—A Brief History: The Warsaw Lectures*. Baylor University Press, 2010.

Young, Iris. "The Logic of Masculinist Protection: Reflections on the Current Security State." *Signs*, vol. 29 no. 1, 2003, pp. 1–25.

Zahn, Gordon. "War and Its Conventions." *Worldview*, vol. 16 no. 7, 1973, pp. 25–33.

Zinn, Howard. "A Just Cause, Not a Just War." *The Progressive*, December 2001, https://progressive.org/magazine/just-cause-just-war-Zinn/.

Zornberg, Avivah Gottlieb. *The Beginning of Desire: Reflections on Genesis*. Random House LLC, 2011.

Index

absolutism 110–121
Addams, Jane 51, 64–65, 111–112, 175–180
alcohol 175–181
alienated war 154–159
anti-colonialist pacifism 47–50
anti-imperialist pacifism 45–47
appraising 81–84; personal pacifism 86–89
appropriateness 118–119
autonomous logic 9–11

"big tent" pacifism 43–44
blindness 21–22

Clausewitz, Carl 10, 92–95, 121
Cold War: after 154–159; and martial liberalism 29–31; redux 37–40
complexities, historical 183–188
confronting the war system 8–13
crisis, time of 5–6
critical war theory 76–80
critiquing the war system 8–13

defection 17–19
deficits 138–143
desolation 138–143
despair 176–180
dialogue, pacifism 62–65
disarming peace 159–162
disarming power 163–165; alcohol, heart-broken women, and hope 175–181; excitement and sacrifice 171–175; repentance before the Flood 165–171
disillusionment 17–19
drill 138–143

dynamic of the war system: after the Cold War 154–159; disarming peace 159–162; the global system 127–136; medieval frameworks and modern cycles 122–127; stages of the war system 136–154
dynastic state 138–143

empire building 95–101
empire, emergence of 131–133
empires 98–101; free trade empires 143–148
excitement 171–175
extrusion 125–127

feminist pacifism 50–52
feudalism 122–125
free trade empires 143–148

Gandhi, Mohandes 191–193
global system 127–136
"good" war, the 25–29
Grand Disillusionment 17, 31–33
Grand Illusions: challenges to pacifism 15–20; pacifism as war abolitionism 7–15; spectacle of war 20–40; violent age 1–7
grief 163–165; alcohol, heart-broken women, and hope 175–181; excitement and sacrifice 171–175; repentance before the Flood 165–171

heart-broken women 175–181
historical complexities 183–188
historical dynamic: logic of war 11

hope 163–165; alcohol, heart-broken women, and 175–181; excitement and sacrifice 171–175; repentance before the Flood 165–171

imperial fantasies 23–25
implosion 125–127
inexorability 10–11
inhumanity 104–105
injustice 104–105
institutions of war building: and inexorability 11
isolation 176–180

killing: personal pacifism's appraisal of 86–89
King, Jr., Dr. Martin Luther 191–193

logic of war: contradiction 12–13; historical dynamic 11
love 180–181

maligned, pacifism as 16–17
marginalized, pacifism as 16–17
martial liberalism 29–31, 151–154
medieval frameworks 122–127
memory 163–165; alcohol, heart-broken women, and hope 175–181; excitement and sacrifice 171–175; repentance before the Flood 165–171
modern cycles 122–127
moral idiot, pacifist as 67–69
Mumford, Lewis 129, 169–171

Narveson, Jan 190–191
nation-states 143–148
"New Wars" 154–159
9/11 33–37
nonresistance 52–58
nonviolence 58–62

offensive, assessing wars as 119
opposing/opposition 81–84; personal pacifism 89–90
optimism 21–22

pacifism 13–14; challenges to 15–20; personal and political 81–121; and self-defense 182–193; as too categorical 114–116; as too

presumptuous 116–119; as tradition 42–80; as war abolitionism 7–15
"Pacifism as Pathology" (Churchill) 65–67
paradox 125–127
peace, disarming 159–162
peace building 106–109
peace making 106–109
personal pacifism 84–86; appraisal of killing 86–89; opposition to war 89–90
personal self-defense 188–189
philosophy see political philosophy
political pacifism 90–92; confronting the war system 106–110; critique of the war system 101–105; structure of the war system 92–101
political philosophy 5–6
post-World War II 151–154
power, disarming 163–165; alcohol, heart-broken women, and hope 175–181; excitement and sacrifice 171–175; repentance before the Flood 165–171
power, reconceiving 109–110

rational 117–118
repentance 165–171, 179
resistance 125–127
revisionism 73–76

sacrifice 163–165, 171–175; alcohol, heart-broken women, and hope 175–181; excitement and sacrifice 171–175; repentance before the Flood 165–171
self-defense 182–183; Gandhi and Dr. King 191–193; historical complexities 183–188; Narvson's critique 190–191; war and personal self-defense 188–189
slam-dunk arguments 14–15
spectacle of war 20–40; after 9/11 33–37; Cold War and martial liberalism 29–31; Cold War redux 37–40; the "good war" 25–29; grand disillusionment 31–33; imperial fantasies 23–25; optimism and blindness 21–22; violent 20th century 22–23

state, the: dynastic 138–143; emergence of 128–131
state building 95–101
state–empires 23, 98, 133–136
states 95–98; nation-states 143–148

terrorism 154–159
total war 143–154
tradition: anti-colonialist pacifism 47–50; anti-imperialist pacifism 45–47; "big tent" pacifism 43–44; critical war theory 76–80; feminist pacifism 50–52; nonresistance 52–58; nonviolence 58–62; pacifism as dialogue 62–65; pacifism as "pathology" 65–67; pacifism(s) 42–65; pacifist as moral idiot 67–70; revisionism 73–76; Walzer's revival 70–73

violent age 1–7, 22–23

Walzer, Michael 70–73
war and personal self-defense 188–189
war, alienated 154–159
war, logic of: contradiction 12–13; historical dynamic 11

war, opposition to: personal pacifism 89–90
war, spectacle of 20–40; after 9/11 33–37; Cold War and martial liberalism 29–31; Cold War redux 37–40; the "good war" 25–29; grand disillusionment 31–33; imperial fantasies 23–25; optimism and blindness 21–22; violent 20th century 22–23
war, total 143–154
war abolitionism: pacifism as 7–15
war as a practice 119–121
war building: critique of 101–104; and inexorability 11; and personal pacifism 92–95
war making 92–95; critique of 101–104
war system: after the Cold War 154–159; confronting 8–13, 106–110; critique of 8–13, 101–105; disarming peace 159–162; the global system 127–136; medieval frameworks and modern cycles 122–127; stages of 136–154; structure of 92–101
Westphalian Settlement 133–136
women, heart-broken 175–181

Printed in the United States
by Baker & Taylor Publisher Services